Sociological and Medical Aspects of Nutrition

World Review of Nutrition and Dietetics

Vol. 55

Series Editor
Geoffrey H. Bourne, Grenada, West Indies

KARGER

Basel · München · Paris · London · New York · New Delhi · Singapore · Tokyo · Sydney

Sociological and Medical Aspects of Nutrition

Volume Editor
Geoffrey H. Bourne
St. Georges University School of Medicine, Grenada, West Indies

16 figures and 39 tables, 1988

KARGER

Basel · München · Paris · London · New York · New Delhi · Singapore · Tokyo · Sydney

World Review of Nutrition and Dietetics

Library of Congress Cataloging-in-Publication Data
Sociological and medical aspects of nutrition.
(World review of nutrition and dietetics; vol. 55)
Includes bibliographies and index.
1. Nutrition. 2. Nutrition--Social aspects. 3. Food habits. 4. Nutrition disorders.
5. Nutritionally induced diseases. I. Bourne, Geoffrey H. (Geoffrey Howard), 1909-.
II. Series.
[DNLM: 1. Nutrition. 2. Nutrition Disorders--complications. 3. Socioeconomic Factors.
W1 W0898 v. 55 / QU 145 S678]
QP141.A1W59 vol. 55 612.3 s [613.2] 87-26094
ISBN 3-8055-4703-X

Drug Dosage

The authors and the publisher have exerted every effort to ensure that drug selection and dosage set forth in this text are in accord with current recommendations and practice at the time of publication. However, in view of ongoing research, changes in government regulations, and the constant flow of information relating to drug therapy and drug reactions, the reader is urged to check the package insert for each drug for any change in indications and dosage and for added warnings and precautions. This is particularly important when the recommended agent is a new and/or infrequently employed drug.

Printed in Switzerland by Thür AG Offsetdruck, Pratteln
ISBN 3-8055-4703-X

Advisory Board

Contents

Possible Nutrient Mediators in Psoriasis and Seborrheic Dermatitis

I. Prevalence, Etiology, Symptomatology, Histological and Biochemical Features

Possible Nutrient Mediators in Psoriasis and Seborrheic Dermatitis

II. Nutrient Mediators: Essential Fatty Acids; Vitamins A, E and D; Vitamins B_1, B_2, B_6, Niacin and Biotin; Vitamin C Selenium; Zinc; Iron

Wld Rev. Nutr. Diet., vol. 55, pp. 1–40 (Karger, Basel 1988)

Sociological and Social Anthropological Approaches to Food and Eating

Anne Murcott

Department of Sociology, University College Cardiff and
Department of Psychological Medicine, University of Wales College of Medicine, Cardiff, Wales, UK

Contents

Introduction

Social anthropology has paid considerably more attention to the study of food and eating than has sociology. Indeed its scholarly interest in food is probably as old as the discipline itself [e.g. 45, 71, 99]. Anthropological studies specifically focussed on food received a boost in the 1930s with Audrey Richards' [97, 98] pioneering work and by the 1940s Margaret Mead had also turned her attention in this direction [73–77]. The period since then has seen a steady increase such that among other developments nutritional anthropology [15, 43, 58, 95] has emerged in the USA as a recognised specialty – one that has, however, no direct counterpart in the UK. At the same time, it is probably fair to say that the great debates in post-war social anthropology revolve around topics other than approaches to the study of food habits. So, for instance, we can still find Robson [101] in 1980 pointing to a neglect of the anthropology of food.

Sociology, on the other hand, has barely recognised the study of eating habits – until, that is, very recently. It is as if sociologists, like the populations whose societies they study, could take a more than adequate supply of food for granted. When sociologists do occasionally consider food it is in precisely those arenas of industrialised society where adequate eating cannot be assumed; during wartime [1], associated with poverty [103, 104, 108], where pathological eating behaviour develops [see 13, 62 for the stigmatising consequences of being considered overweight], 'unorthodox' dietary styles [102] or, most recently in Britain, in the wake of debates about the development of a nation-wide food policy [106, 118]. Such work thus straddles the often uneasy divide between sociology on the one hand and the study of social policy and the application of sociology to society's problems on the other. This tends to keep the topic in view as a variety of social problem [e.g. 3, 9, 109], i.e. as part of just one, albeit vitally important corner of the sociological enterprise.

What follows is not a review in the usual sense of the word – if by review, we mean an article that brings together summaries of research findings and assessments of research methods. Rather (with a multi-disciplinary readership in mind) it is an introduction to and illustration of some of the ways social anthropologists and sociologists have studied food and eating, their analytic approaches and theoretical viewpoints. It deals centrally with the beginnings of a dialogue about such approaches and viewpoints.

The disciplines of social anthropology and sociology are treated together here. Despite their obvious and familiar differences, they are close neighbours in their general intellectual inheritance, and in the fact that each is less easily characterised as a unitary discipline than, plurally, as consisting of a number of theoretical perspectives. More specifically, they are close in, among other things, the insistence that any aspect of human activity has to be studied in relation to the social context in which it occurs. It means, then, that the discussion to follow ranges well beyond an ostensible focus on food and eating.

In keeping with this broad substantive concern, the term 'food habits' is used in this review in a general rather than precise sense. Previous attempts have been made to specify terminology – e.g. Simoons' [107] 'foodways', Wellin's [113, 114] 'food ideologies', Douglas' [30] 'food event' – and to introduce standardisation [65]. But the relative lack of detailed development of the field possibly makes satisfactory standardisation premature. In addition, the use here of 'habit' is not intended exclusively to refer to behaviour that is 'merely' habitual, i.e. somehow unthinking or accountable by reference to a notion of tradition. 'Food habits' is used here merely as a provisional, convenient and inclusive shorthand to cover the widest possible range of food choice, preferences, meal patterns and cuisines.

That said, however, the focus in this review is on food rather than drink. Use/abuse of alcoholic drink is, for present purposes, excluded (but see MacAndrew and Edgerton's [70] compelling study offering a sociological/anthropological explanation of drunkenness cross-culturally; for an introduction to the sociology of the genesis of alcoholism as a medical problem see Robinson [100], and cf. also the 'classic' study of pre-war British social drinking by Mass Observation [72]).

This review is selective in a number of ways. First it covers only the work of those sociologists and social anthropologists working in a professional academic capacity. Second, it does not deal with the work of other scientists interested in social aspects of food habits, such as social nutritionists [10, 93, 94, 122] (their field has developed more as a branch of the nutritional than social sciences) or social historians [16, 44, 59]. Third, it concentrates on the last decade or so and looks only at literature readily available in English.

Although attention to food and eating has been somewhat marginal to the mainstream of work in social anthropology and especially in sociology, a reasonable number of studies has been published in the last 25 or so years [e.g. 14, 17–20, 22, 38, 46, 60, 84, 92, 110–112, 116, 119, 121]. Such studies could be said to have a common starting point which (though it is not always spelled out) can be summarised in the following four steps: (1) human beings are omnivores; (2) yet they do not eat everything available that is potentially nutritious; (3) this is only partly explainable in biological terms; (4) thus there have to be cultural bases for the selection people make.

Beyond having this basic assumption in common, however, these studies diverge considerably in terms of approach, method, and selection of substantive research topic. Indeed, taken as a whole, this literature appears inchoate. The reader is left with the impression that scholars more often than not seem to be working pretty much in isolation from, even ignorance of, one another's efforts. Even though some edited collections, reviews and general texts [4, 21, 35, 36, 86] have been published, they only very partially counteract the impression of fragmentation and underdevelopment. Above all, there is little sense of explicitly theoretically informed approaches, or of debates and dialogues. Considered discussion about *how* the cultural bases for food habits might be studied and what style of explanation shall be adopted to account for them has been almost wholly absent from this literature – until, that is, very recently.

For a dialogue about the merits of different theoretical approaches that draws on existing general debates in both disciplines is now beginning to surface *within* the literature that deals with the sociology and anthropology of food habits. This, then, is the development with which the present review is concerned. It can be outlined as follows.

The first major thesis in this debate within the food habits literature, that of the *structuralists,* was put forward in the 1960s and had achieved a dominance in those parts of the literature that were theoretically considered by the end of the next decade. Such an approach is concerned with the significance of the symbolic aspect of cultures. It singles out the way we can, in the florist's slogan, 'say it with flowers' – except that given the present concern, it is the way we say it with food. In the USA people say Thanksgiving with roast turkey, and in parts of Africa they nowadays may say weddings with fruit cake coated in white royal icing [117].

Into the 1980s, however, a new group of writers has appeared in the field of food habits: a group who draw on another, equally well-established tradition of analysis in the social sciences, but one that complements or sometimes rivals the symbolic approach. In this alternative stance, the stress is on the practical rather than the symbolic, on the economic and the political rather than the purely communicative and ideational. While not altogether denying that we might speak as we eat, these critics point out that life does not exclusively consist of self-expression, for there are all sorts of practical limitations and interruptions that shape the possibility of such expression – European colonial expansion brought most Americans to the turkey and the wedding cake to Africa.

The rest of this paper examines the debate between these two very different points of view, looking at relatively few scholarly contributions in some depth; on the one hand the structural approach as evidenced particularly by the work of two anthropologists, Douglas [23–31] and Lévi-Strauss [66–69]: on the other hand, the stress on the practical, economic and political aspects of food habits to be found in four books published within the last 4 years by Harris [55], Mintz [82], Goody [51] and Mennell [78] – respectively three more anthropologists and a sociologist. We begin with the structuralists.

Structural Approaches – into the 1970s

Preamble

A structural approach represents a specialised view of the insistence that it is to culture we must look to explain the variety of human food habits. At its heart, the approach acknowledges the exquisitely human capacity both to deal in symbols and to confer symbolic significance on objects that surround us. Eating habits illustrate this as vividly as other forms of human activity – perhaps especially so (for a 'light read' exemplifying this approach to food habits see Murcott [88]). As we eat we not only survive corporally, we also express ourselves socially. We all need nourishment, and from the nutritionist's point of view, some foods are much like many others. Lentils, steak, duck eggs or oysters are virtually interchangeable as sources of protein – their nutritional value is then to be assessed in terms of amounts taken and in terms of their combination with sources of other necessary nutrients. Socially, however, the items just mentioned can familiarly have widely contrasting connotations – of status, of wealth or of occasion.

Nutritional values are one thing, social values quite another. People attribute meanings whose origins are essentially social to external objects, events, circumstances – and also one another – in a way that is familiar and makes sense to them. Study of the symbolic significance attached to foods can thus demonstrate the human ability to construct a world of ideas that imbues the material environment with meaning. A preoccupation with such matters is the starting point of a structural approach.

Before delving further into this perspective, two preliminaries must be mentioned. First, singling out the work of Douglas and Lévi-Strauss is not to overlook the contribution in this vein of the latter's compatriots; for instance Barthes [11] has dealt with matters of taste, in fashion, as well as food. Second, it is merely for the sake of convenience that I refer to this approach as structural. The thought of Lévi-Strauss is widely identified as *structuralist,* yet at the same time Douglas will describe both his and her own approach as *structural.* In turn, others [15, 120] describe her anthropology as cultural. Too much weight should not be placed on these terms; and in this review I shall talk of 'structural approaches' or 'structuralism(s)' interchangeably, simply for the sake of a more tolerably readable account. In any case, there are critical differences between Douglas' and Lévi-Strauss' stances, as we shall see, turning first to Lévi-Strauss.

Lévi-Strauss

A good deal of the French anthropologist Claude Lévi-Strauss' extensive work concerns food and eating [66–69]. See also Leach [64] for an accessible introduction both to Lévi-Strauss' work on food and his thought more generally, and cf. Leach [63]. And for much of that, his attention focusses on cooking, though his approach to the topic has to be seen as part and parcel of his whole anthropology.

Cooking is a universal and distinctively human activity. No matter how diverse food choice may be worldwide, all known people transform much of what they eat by heating it in one or other fashion. Unlike animals which, with rare exceptions, eat their food in the state they capture or come across it, human beings not only wash, wipe or peel their food, not only preserve it by drying, cooling or smoking, routinely they also cook it.

It is Lévi-Strauss' enduring concern with universal features of the human species that leads him to take a close interest in such familiar facts of life. What, he asks, is the significance of cooking as a peculiarity of the species? His answer points to a paradox. We all know that human beings have a dual character. As a species we are animal, more or less like other animals; we have parallel bodily functions and similar biological requirements for survival. At the same time, however, we are not animal but human. We have language, intelligence and culture blended in a way that is unique on this planet. Thus we are creatures of 'Nature' and 'Culture' simultaneously.

Moreover, knowledge of this universal feature of human beings is itself universal and forms part of the cultural awareness of all societies. Yet, so he argues, it is also a source of latent tension in human affairs which perennially demands resolution. Like all good paradoxes, however, it defies once and for all solution. In Lévi-Strauss' view, the need to eat is one way we are repeatedly faced with the paradox, and cooking one of the means whereby we continually express our attempts to cope with it.

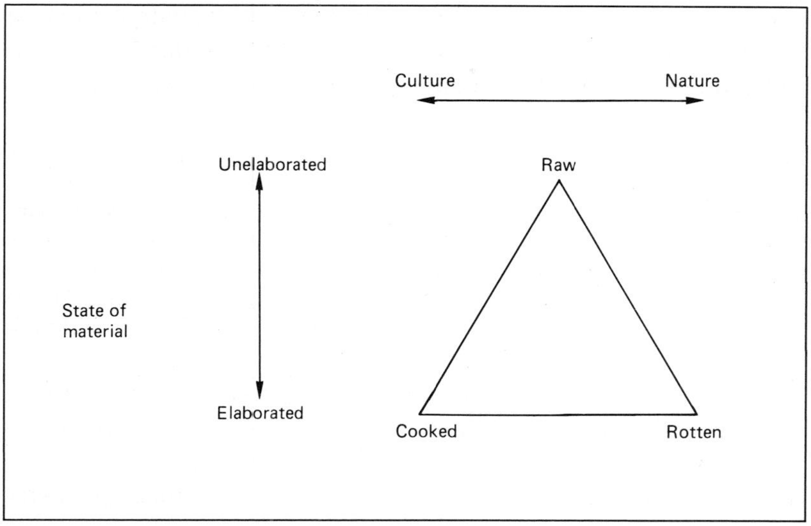

Fig. 1. The culinary triangle. After Lévi-Strauss [66].

The most accessible instance of his approach is to be found in a short essay [66] which was first published the year after *The Raw and the Cooked* [67] (volume I of *Mythologiques* whose first three volumes in particular involve consideration of the human significance of various transformations of foodstuffs). Here he links the human universal in which he is interested, i.e. cooking, with another human universal, language. Not only are they similar by virtue of their universality, they can also be linked, he claims, by applying the methods for exploring one to examine the other: analysing the structure of language can be used to reveal the structures of culinary activity. So doing is justified by his observation that each is a form of essentially human expression.

Borrowing, then, from a branch of linguistics, Lévi-Strauss proposes a scheme he calls the culinary triangle (fig. 1). Its basic building blocks consist simply of a pair of binary oppositions. These he imaginatively derives from an obvious point about cooking: heating food entails a transformation, an elaboration, of the original, raw ingredients. Thus we immediately have their binary opposition, transformed/normal (or marked/unmarked, *elaboré/non elaboré*). But cooking is only one way in which what might be used as ingredients can be transformed: they all too obviously can go off. In other words, Nature can transform them by process of putrefaction. And so Lévi-Strauss produces his second binary opposition, that between Nature and Culture. Put this double opposition together and the result is

> a triangular semantic field whose three points correspond respectively to the categories of the raw, the cooked and the rotted. It is clear that in respect to cooking the raw constitutes the unmarked pole, while the other two poles are strongly marked, but in different directions: indeed, the cooked is a cultural transformation of the raw, whereas the rotted is a natural transformation. Underlying our original triangle, there is hence a double opposition between *elaborated/unelaborated* on the one hand, and *culture/nature* on the other.
> (Lévi-Strauss [66], p. 937; italics in the original.)

Lévi-Strauss then develops this scheme to take account of different methods of food preparation and preservation (roasting, smoking, boiling and so on) and constructs a developed form of the triangle to do so (fig. 2). He even goes on to indicate the possibility of more complex geometry that incorporates additional dimensions.

By this stage, some academics' credulity has been stretched too far. Mennell [78, p. 9] calls the triangle a 'farrago of nonsense', and even so sympathetic an exponent as Leach [64, p. 32] describes it as a 'game of acrostics ... this fandangle'. Why, then, give the culinary triangle so much house room in this review? For the simple reason that however unlikely parts of his argument undoubtedly are, Lévi-Strauss seems to be on to something.

In stressing the dual character of human beings, he reminds us that unlike animals whose food choice and eating behaviour is governed by instinct, once weaned human beings have to learn what to do. And what they learn are the cultural conventions that decree what is to count as food, which foods are appropriate for what occasions and so on. And since occasions are identified socially, then, a structuralist view proposes there must be some patterned correspondence between kinds of food and types of occasion.

James' [57] work provides an illustration of the point. She uncovered an unexpected pair of meanings to the same N.E. English dialect word 'ket'. Adults use it to mean rubbish, something worthless, children use it to refer to the type of sweets they, not adults, choose to eat. James [57] develops an analysis first of the significance of these sweets showing that, in metaphoric terms, they break the rules of edibility. For they come in garish, i.e. non-food, colours, taste acrid or gritty and are named or modelled to represent the literally inedible (such as telephones or car parks) or the

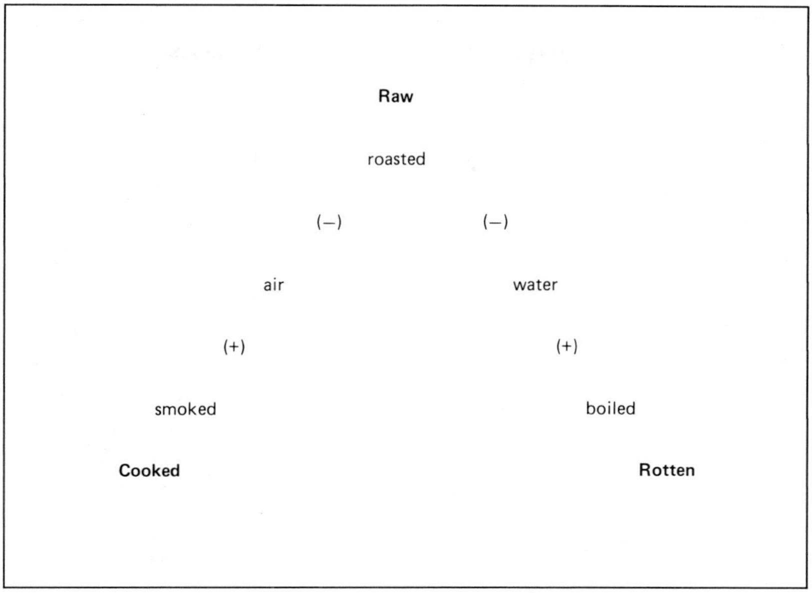

Fig. 2. The culinary triangle developed. After Lévi-Strauss [66].

proscribed (gorillas, cats or footballers). Then she records the context of children's eating behaviour and patterns, noting the direct contrast with adult manners and parental emphasis on regular meals that are not to be spoiled by eating sweets in between times. And, finally, she proposes that when children prefer such seemingly unlikely eating they are expressing a relatively harmless form of rebellion by consuming what adults call rubbish.

Lévi-Strauss' work can take us further. Though the availability of foodstuffs varies in different parts of the world, there are nonetheless recognisable similarities in the way people of obviously different cultures think about food. All people classify what they eat, subdividing the general category 'food' into sub-groups to be prepared or eaten in different ways. And social use has its symbolic aspect. In well-off industrialised countries, meat can take pride of place on a menu with potatoes as accompaniment; roast chicken is for Sundays while the leftovers go into Monday's chicken pie as the working week begins again. We are quite used to such ways that items are accorded different valuations in relation to one another in terms of some hierarchy of prestige. But Lévi-Strauss has marshalled a good deal of evidence to show that so too are people with cultures – and foodstuffs – very different from our own. What is more, he claims that common structures can be detected in categorisations of foods and classifications of their use, and that, moreover, there are universal principles underlying those structures. He sums up his point in a phrase referred to by both Mintz [81, 82] and Harris [53–55] (whose work will be considered later). Food is not so much 'good to eat', but rather 'good to think'.

Sorting out the worthwhile from the far-fetched in Lévi-Strauss is, however, no easy matter. Whether in French or English understanding his writings is difficult to the point of obscurity. And this is the least of the problems his critics identify. Objections to his work range from scepticism about a method that relies on the contrivance of binary oppositions [29, 51], accusations of failing to practise the analytic principles he himself preaches [64, 78] and charges that he is careless, even cavalier in his marshalling of the data [35]. And though hailed by Leach [64] as the most distinguished exponent of social anthropology anywhere outside the English-speaking world. Lévi-Strauss practises in quite a different tradition from Anglo-Saxon varieties. His ultimate concern to establish facts about the 'human mind' is foreign to English-speaking social anthropological attention to the social organisation of societies.

Nevertheless, even critics credit Lévi-Strauss with what is probably the most extended effort to understand the symbolic aspect of food habits – an aspect that simply cannot be ignored. So his work, though at times bizarre, represents a serious attempt to answer the key question about which style of explanation we should adopt to study human food habits. Others who also take this question seriously find they cannot completely write him off.

Douglas

Initially based in Britain then in recent years in the USA, Mary Douglas has sustained a constant interest in the social significance of food and eating. Her writings specifically on the topic span something like 20 years, ranging, to name only some, from an analysis of the dietary laws of Leviticus [23], the sequencing of meals and menus [24, 30] to the promotion of research programmes [26] and the co-ordination and editing of studies of food in American communities [29].

Her interest has, however, to be seen as more extensive still. For her, the business of food and eating is of a piece with a wider range of topics with which she has been concerned. What is more, as with Lévi-Strauss, her perspective on food habits has to be understood as an integral part of her whole anthropological approach. Her structural perspective can perhaps be best appreciated by bearing an example in mind.

Consider the instance of some food left on a person's plate after they have had enough to eat. These scraps may be given to the dog, or just thrown in the dustbin. In the ordinary way such leftovers would not be set aside to be served again at another meal. Nor do those at the table usually finish what each other leaves: well-behaved people do not eat off one another's plates. Such familiar arrangements of modern life would seem to be justified in terms of hygiene as well as good manners.

But how come rules of hygiene are set aside when lovers share an apple? or good manners abandoned as a mother affronted by the waste of good food she prepared herself finishes what her 2-year-old refuses to eat? And why does care about hygiene apply selectively such that black teenagers claim that their white comrades only wipe the mouth of the bottle of Coca Cola the whole group is sharing when a black not a white friend has taken the preceding swig?

This may be a slightly sideways look at food habits. But it serves to show how Douglas' [23] structural approach to the matter locks into place alongside her earlier work on ritual, taboo and pollution and her later focus on the world of goods, consumption in its many aspects [28, 31]. First it exemplifies a cardinal principle that we can start to understand food habits only in relation to the cultural context in which they occur. Research into the cultural aspect of food, she says,

> should avoid the mistake of isolating one activity from the whole. The food system is only one of a number of family systems for caring for the body ... (and) the culinary tradition is only part of the total cultural experience of an individual. (Douglas [26], pp. 74, 79.)

Once we see this, then we can gain a purchase on the nature of the meanings attributed to food, amongst many other things. Food is not food in itself, but becomes identified as such in relation to a whole system of ideas. What is left on the plate is regarded as inedible only by virtue of its being there: what was safe and clean to eat at the start of the meal has not that long after become unclean and inappropriate. Such an angle on the question is of a piece with Douglas' [23] analysis of dirt as no more than 'matter out of place'.

> It is a relative idea. Shoes are not dirty in themselves, but it is dirty to place them on the dining table; food is not dirty in itself, but it is dirty to leave cooking utensils in the bedroom, or food bespattered on clothing ... (Douglas [23], p. 48 of 1970 ed.)

Douglas pursues this line of reasoning much further. Food is used to signify not just people's relation to objects, but also to their own bodies. It displays something of our ideas about what is acceptable and safe to take into the body – it betrays the meanings we attach to our bodies in their objective aspect. And our use of food reveals into the bargain concepts of closeness and acceptability in our relationships with one another. Different rules of hygiene can apply between partners or parents. Dogs and the occasional vagrant who can be seen foraging for food in trash cans, that eat leftovers, are (like it or not) classified together as lower orders. (For a historical account of the use of leftovers, see Aron [5].) And even in a multi-ethnic group of friends, deep-seated prejudice and avoidance can be displayed in the fleeting but telling instance mentioned above. Indeed, the very fact that wiping the mouth of a shared Coke bottle is, to many of us, hygienically dubious anyway only serves to underline the metaphoric, the symbolic nature of the gesture.

Like Lévi-Strauss, Douglas sees human beings 'speaking' as they go about their daily lives. Everyday life is remarkable for its orderliness and system in no matter which culture. For Douglas, this order permeates all levels of society right down to the minutiae of bodily routines and rituals. Moreover, the order, the structure, at every level echoes and is articulated with every other. Food, the composition of menus, the succession of meals constitute a medium of communication. Their nature and structure encodes the further structures in terms of which those other aspects of society's organisation are classified.

Despite many similarities, Douglas' structural perspective must not be confused with Lévi-Strauss' brand of structuralism. Her work is firmly in an Anglo-Saxon anthropological tradition in modes of analysis and approach to evidence. Her fiercest criticism of his stance is probably her objection to his reliance on binary oppositions. To Douglas their derivation is quite out of keeping with her insistence that the symbolic structure has to fit squarely with some demonstrable social consideration in the whole complex of cultural life under scrutiny. To do otherwise is invalid, and the analysis of symbols becomes no more than arbitrary and subjective.

Two further points about Douglas' work on food must be mentioned. First, in addition to analyses of text, Douglas has also undertaken fieldwork in modern society. With Nicod [30, 90, 91] as researcher she undertook a small but intriguing study of day-to-day eating habits in Britain. The most interesting aspect of their study is the method – Nicod closely observed the eating habits of four families with whom he lived in turn as a lodger. This strategy ingeniously got round problems of recall involved in asking questions, and sought to keep observer-effect to a minimum.

The result, however, is not wholly satisfactory. Representativeness can quite legitimately be traded off against the collection of data in depth in a small number of cases. But unfortunately, in this study the loss of the former is not convincingly compensated for by the latter. Moreover, full enough attention to Douglas' own precepts to use analytic categories that derive from the situation under scrutiny is missing. This limitation does not, however, invalidate the effort. Rather it provides a most important lead – one which so far, it seems, no one has followed.

Lastly, unlike the other main authors considered in this review, Douglas seeks to address the detailed practical concerns of nutritionists and dieticians (though it must be said that it is often somewhat allusive). It is no good, she says, trying to change people's eating habits without understanding that there is a system in the structure of each meal and in the pattern of meals overall. It is a system to which people are attached because it is a medium whereby they routinely communicate a variety of matters of fundamental importance to them in making sense of the world, and thereby reassure themselves that 'all's right' with that world. And those attachments are stronger at some parts of the system than others. Thus if changes are to be introduced, they are most likely to be successful by learning which are the weak points and starting from there.

In all, Douglas' work represents one of the most outstanding contributions to the anthropology and sociology of food habits. Despite occasional unevenness, she offers some brilliant analytic insights in her attention to the minutiae of daily life and the place of food and eating within it. What is more, in rejecting the arbitrariness of Lévi-Straussian binary analysis, she provides a far more plausible and appealling variety of structural approach.

Reactions

Lévi-Strauss and Douglas both have their disciples among those studying food habits. In France, for instance, a structural inheritance is evident in the work of Fischler [39–42]. And in Britain, Douglas, as well as Lévi-Strauss, provides an inspiration for the next generation introduced to her work as students. In addition to James' [57] work already mentioned Atkinson [6–8] has developed an imaginative analysis of modern fashions for whole foods, 'alternative' dietary remedies and moral pressures on eating habits more generally (see also Murcott [83]).

Yet structural approaches, whether the one adopted by Lévi-Strauss or the other favoured by Douglas, have serious limitations. In addition to the severe doubts about some of Lévi-Strauss' analytic tactics, both he and Douglas (with the important exception of her work with Nicod) tend to rely, in their discussion of modern eating habits at least, simply on their own experience, or on anecdotes. Clearly, though this may do for a beginning, it will hardly suffice for a developed analysis. Moreover, there are further, more fundamental difficulties, which in varying measures, the four authors, among others, who take us into the 1980s all point to.

One of these is the great risk run by the structuralists' profusion of classificatory schemes when coupled with such lapses in empirical rigour. Classification is fundamental to all science, and structural approaches have certainly proposed some provocative types. Unless, however, due attention is paid to the development and testing of rules by which such classification may be reliably developed, any particular scheme, no matter how potentially useful, can rapidly become merely arbitrary – something imposed by whim or fiat which distracts as much or more than it enlightens. While such a procedure may well provide valuable imaginative insights, it risks no more than restating one set of codes of meaning in terms of another. This circularity represents a serious drawback and is no substitute for explanation.

Another major problem with Lévi-Strauss' brand of structuralism is that it leaves the impression that the search for universal principles in human affairs is the only worthwhile pursuit for the social scientist – a position already noted that is not generally shared in the Anglo-Saxon tradition. There is, then, a static quality built in to the style of analysis. It conveys a sense of timelessness in leaving any question of historical change out of the picture altogether.

Finally, the structural viewpoint is accused of giving undue weight to the symbolic sphere and neglecting reasonable consideration of material factors in seeking to understand human food habits. Eating after all is also a practical and indeed essential business. It is all very well to examine human ideas about Culture and Nature expressed via food, but we can also take an opposite (if philistine) view that Nature carries on regardless of what human beings think they are up to. More to the point, we can extend that view to explore the ways human beings *modify* nature, creating social, political and economic systems whose consequences can shape, or at least intersect with, the sphere of the symbolic. So it is to some responses to structural approaches that we now turn.

Materialist Responses – into the 1980s

Preamble

Structural approaches, then, have their drawbacks. Their concentration on the symbolic, on codes and on meanings holds attention on the realm of ideas. To some, such a philosophy verges on a variety of 'idealism' (or 'mentalism') whereby we are asked to see reality as apprehended in thought. The four reactions to structuralisms to be looked at in detail in the remainder of this review, adopt instead, versions of what can roughly be described as 'social materialism', or simply 'materialism'. This alternative philosophy is to be seen as a kind of 'holding operation' against idealism that highlights the practical facts of life.

Generalising in this way overstates matters (Lévi-Strauss, for instance, repudiates the charge of idealism: he implies that there is a biological basis for the universality of the types of classification he identifies) but it does provide a convenient shorthand with which to characterise an essence of the responses to structural approaches. We begin with the work of Harris.

Harris

Like Douglas, the American anthropologist Marvin Harris has discussed the social significance of human food habits at various points in his very considerably body of work [notably 53, 54]. But he takes a diametrically opposed view of the topic from Douglas and others adopting a structural approach. Indeed, his most recent book [55], devoted this time exclusively to food, represents a counterblast against what he considers a runaway idealist approach. And it may not be too fanciful to think that in giving it the title *Good To Eat* he is deliberately spotlighting the part of Lévi-Strauss' dictum (that food is good to think rather than good to eat) which structural approaches play down. His philosophy, expressed throughout the book with a down-to-earth, no-nonsense air about it, is this:

> (T)here are generally good and sufficient practical reasons for why people do what they do, and food is no exception. (Harris [55], p. 14.)

Harris devotes the book to a test of this proposition by taking a number of cases illustrating puzzles of food choice. These range from milk-lovers and milk-haters (his shorthand) and the question of lactose intolerance, Jewish and Muslim avoidance of pork, Hindu avoidance of eating sacred cattle, discussion of which cultures count which animals as pets and whether they are acceptable as food, insects as delicacy or objects of disgust, right along to preferences for or against what he calls the non emergency consumption of human flesh.

We can get an idea of his line of theorising by taking a look at one of these instances in a chapter he calls *Meat Hunger*. Harris begins this case study with the demand for meat in Poland. Not only are people wearily willing to queue for hours in butchers' shops, the government cannot ignore the demand: a cut in meat subsidy in 1981, reports Harris, led to the need to declare martial law to restore order. Why, Harris wants to know, does the government not point out that the population is in no danger of malnutrition, but instead, meets each demand with promises of improved supplies of meat that put the rest of the economy under increasing strain? Is this version of meat hunger merely a symbol of Poles' rejection of their regime's variety of state socialism which the government cannot afford to ignore? To argue as much, says Harris, is to do an injustice to the Polish people. For he believes that eating animal as well as plant foods bestows health and well-being greater than mere survival.

To support this contention, Harris disposes of vegetarianism by dint of counting veganism as the true version (this he asserts is sporadic, unpopular and short-lived) and goes on to a succession of examples of the high value placed on meat eating in cultures as diverse as the !Kung bushmen, the Sharanahua of eastern Peru, the indigenous peoples of Australia and the south Pacific, modern America as well as the Poles.

Such extremely widespread preference for meat, claims Harris, cannot possibly be coincidental. Animal foods must therefore play a special part in the nutritional physiology of our species. In so saying, he is careful to make sure we do not run away with the idea that he is proposing a simple matter of some sort of genetic programming that directs us to meat in the same way instinct leads carnivores such as lions or eagles to their prey. What he means is that our species-given physiology and digestive processes provide the predisposition whereby we learn to prefer animal foods.

At this point in his discussion, Harris turns to the nutritional evidence to remind us of the concentrated goodness – not only proteins but also minerals and vitamins – of animal food sources. And he does not forget their drawbacks. They need to be supplemented with sources of fibre, and the cholesterol story, though incomplete, cannot be discounted. Indeed, he is sceptical of the indictment of meat in the aetiology of cancers and heart disease. Instead he favours the view that increased longevity is a more likely concomitant – and that what is more, meat eating is itself correlated with increasing longevity.

Granting that correlation is not of course evidence of causality, he nonetheless proposes that provided an excess of fats and cholesterol is removed, we should continue to reap the benefits of high levels of animal food consumption. World hunger – as he puts it – for more meat, fish and fowl represents what to Harris is a thoroughly rational preference that arises from the interaction of human biology and the nutritive composition of alternative foods.

With this instance in mind we can now look a bit more closely at Harris' theoretical approach. It goes something like this: what is good to eat (he is still playing on Lévi-Strauss' dictum), i.e. foods that are preferred, are on the whole better than those bad to eat, i.e. avoided, by virtue of offering a more favourable balance of practical benefits over costs. It is a question of weighing up the advantages and disadvantages in three overlapping spheres: the effort and labour involved to acquire food, the comparative costs in these terms of different types of food and the relative advantages in terms of nutritional value of alternatives.

> Even for a omnivore it makes sense not to eat everything that one can digest. Some foods are hardly worth the effort needed to produce and prepare them; some have cheaper and more nutritious substitutes; some can only be eaten at the expense of giving up more advantageous items. Nutritional costs and benefits form a fundamental part of the balance – preferred foods generally pack more energy, proteins, vitamins or minerals per serving than avoided foods. (Harris [55], p. 15.)

His view, incidentally is close to that of ethologists' 'optimal foraging theory', which he cites with approval.

In the case of human beings, however, he is very careful to introduce a couple of warnings. First, he is emphatic that his view is not to be taken as a defence of the status quo. He is not proposing fatalistically that simply because his practical approach can make sense of whatever foodway we may find in the world, that those habits must be best and therefore left undisturbed.

Moreover, just because he is analysing food choices and avoidances in terms of costs and benefits he makes sure that we realise that he does not do so only in monetary terms, and understand that he includes ecological and nutritional considerations in his calculation. To underline his point he draws on the stark and notorious example of the case of selling artificial milk formula for infants in Third World countries. Here the costs were borne by mothers and their babies, and the benefits accrued to the multinational food corporations.

Harris has a number of things to say about his own position vis-à-vis the structuralists. First of all, he proclaims that his own philosophy is currently unpopular. Yet it is hard not to regard this as disingenuous; his type of approach seems to find far more favour with North American nutritional anthropologists [15, 35, 58] than does structuralism of any kind. In any case, his presentation of structural approaches needs comment. For there is a suspicion of mischief in his characterisation of his 'opposition':

> (The) fashionable premise is that foodways are accidents of history which express or convey messages derived from essentially arbitrary values or inexplicable religious beliefs.
> (Harris [55], p. 14.)

He presents a rather watered down version of a symbolic approach which excludes study of the relation of what *appears* arbitrary to other levels of meaning in the social organisation – a key element, after all of structuralisms.

He does, however, go on to pose a significant question: Which comes first, the practical (material) or the symbolic? Harking back to Lévi-Strauss, his answer is:

> that whether (foods) are good or bad to think depends on whether they are good or bad to eat. Food must nourish the collective stomach before it can feed the collective mind.
> (Harris [55], p. 15.)

This is a provocative proposal. But it raises an important analytic question about the nature of the evidence needed to resolve it one way or the other. Very possibly we cannot determine the answer, or if we could, we are unlikely to be able to collect the relevant data. Harris stays silent on the problem, and indeed does not so much debate as simply assert his case.

But perhaps this does not much matter. For the very great value of his work lies in providing a timely reminder for students of food habits of the profound importance of the material. Left virtually to the very end of his book Harris, albeit obliquely, adds an especially telling point. If we are to follow the structural approach, or at least his reading of it, then we risk the very fatalistic acceptance of cultural diversity of which he has been wrongly accused. To accord meanings and messages a higher priority than the material in our sociological and social anthropological analyses may spill over into failing to develop informed ways of tackling the practical problems of the adequate worldwide food provision and distribution facing us all. (His line of argument in this respect is reminiscent of Sahlins' [105], whose theoretical perspective is akin to Harris', but whose work touches on food, and hunger, only as part of his study of 'stone age economics'.)

He would, however, presumably agree that in developing solutions to these problems we must not neglect the importance of the symbolic sphere altogether: for an object lesson on this aspect, see Wellin [115] and also Nichter [89].

Harris' work, then contributes to a dialogue about the way food habits should be studied. Into the bargain he writes extremely fluently and often amusingly – readability is all too rare in serious scientific work. There are, nonetheless, difficulties with the position he outlines in *Good to Eat.* I shall mention three.

To begin with, there is the question of his selection of case studies. At first sight, the range of food choices he considers seems suitably wide. Indeed, he makes play of his claim that he has not selected only those examples which he knew in advance conveniently fitted his analytic approach. Yet it is striking that his selection largely excludes food from other than animal sources. Moreover, he seems studiously to avoid, despite his protestations, what may prove to be the toughest tests of his philosophy, namely learned preferences for elaborate eating among the 18th century European aristocracy, or the ingenious products of the modern food industry.

Certainly types of cost/benefit analysis can be applied to these instances too – e.g. the financial cost to the aristocracy is outweighed by the benefit derived from marking out their superior social status, or the benefit of domestic harmony to a parent outweighs the monetary (and nutritional?) cost of giving the children the largely 'empty calories' of candy bars. But adoption of this line of thinking is, in Harris' own terms, partial – it does violence to his insistence that the calculation of advantages and disadvantages has to include both ecological and nutritional considerations.

Second and most importantly, the assumption at the centre of his analysis, that rationality prevails in weighing up those advantages and disadvantages, is not well supported for individuals, let alone whole cultures. The gamut of social sciences, psychology, sociology and social anthropology and even some parts of economics, all point to the apparently irreducible *irrationality* of human affairs. And even if individuals, or even cultures, *did* act rationally, it is impossible to imagine that the full range of information necessary to arrive at the best answer to a cost/benefit calculation is perfectly available.

There is a further difficulty with Harris' work. Though he draws on historical instances and inheritances, he nonetheless sidesteps study of historical development and changes over time – not least changes in preferences for particular food items themselves, let alone transformations of whole cuisines. His position is not, of course, incompatible with consideration of historical change, and a Harris-style theory of change might be developed. So far, however, he does not appear to have done so. These matters are, however, tackled by the next three authors to be considered.

Mintz

Of the four versions of a materialist response under review here, those of Marvin Harris and his compatriot Sydney Mintz are closest to one another. Both reflect the influence of Marxist thought, though in Mintz' case it is a good deal more obvious. And both regard the material as prior to the symbolic. So it is perhaps, on the face of it at least, all the more surprising to find that *Sweetness and Power* [82] reveals Mintz as the most sympathetic of all four to the insights offered by a structural approach. In this book, Mintz brings some of the subjects of his other writings – the work and organisation of the plantation production of commodities and associated issues such as slavery and colonialism – together with a consideration of aspects of food choice (also cf. Mintz [81]). But he elects to focus on just one instance – a preference for sugar on such a scale that for some countries it is possible to speak of a 'national sweet tooth'.

It is well known that Britain led the way: Mintz cites one report that between 1663 and 1775 consumption in England and Wales increased about 20-fold whereas the population did not quite double in the same period. On another set of comparisons, English consumption increased from around 4 lb per head per annum in 1700 to roughly 108 lb per head by 1972. Nowadays, as Mintz drily observes:

(F)or countries that are already big consumers, like Ireland and England, sucrose may contribute 15–18% of total energy consumption per capita. Allocation of differential quantities by age and class, if we had the data, would reveal a remarkable, if not shocking, dependency of some age/class sectors upon sucrose. (Mintz [82], p. 197.)

There are, of course, some parallels with other commodities, such as tea, coffee, cocoa (and tobacco), but the growth of sugar nonetheless stands out quite dramatically.

Mintz sets himself the task of accounting for so prodigious a development in food choice. Paying careful attention to the evidence for a biological basis for the human liking for sugar, Mintz is nonetheless driven to observe that though some in-built predisposition to sweetness seems unarguable, it cannot possibly explain the extent of degrees of preference and types of taste that we know exist. It really has to be matter of food choice – or so it seems, for no one is literally being force-fed sugar. Thus we must look for other factors sufficiently powerful to explain so massive a demand.

Though Mintz starts his story in the sugar plantations of Puerto Rico where he himself undertook anthropological fieldwork early in his career, he purposely covers a lot of ground historically and geographically. Dividing the body of his study into three main sections he deals first with sugar *production* and details agricultural and manufacturing processes, changes in the technology of refining, developments in the organisation of the work (including the importation of slave labour and the aftermath of abolition) bringing us up to date with a view of modern conditions.

His next long section considers sugar *consumption.* Here he describes the shifts in its use: in the early modern period sugar served sparingly among the very well-off as a medicine and found its way into food on a par with other precious spices from the Far East. But by the 19th century sugar had become a necessity right down to the labouring classes. Conscientiously, Mintz documents sugar's percolation from the wealthy down through the social strata, and its spread from metropolis to hinterland.

It is, however, merely for convenience that Mintz splits the discussion into production and consumption. For he is at pains to make clear that part of his analytic approach depends on recognising that the two go hand in hand. Demand and supply are not to be seen as independent of one another, but are merely shorthand to describe complex forces that contribute to each other's generation. Demand induces supply, and supply lubricates increasing demand.

More important still, the two do not develop independently of human agency and interest – which is where power (part of the book's title) enters the picture. No matter, Mintz says, how knotty a concept it is to define, the power wielded at all levels, from plantation and factory owners, through international traders to the political interests of nation states, cannot be ignored. It is an integral part of the mammoth development in the choice of sugar.

It is into this third major section on power that Mintz weaves insights concerning the symbolic significance of eating. The meanings people attribute to sugar changed dramatically; from those of 17th century England when it signalled luxury and display as the prerogative of the rich and powerful, through those later in the same century when the less well-off came to be able to use it to mark special occasions – the sweetmeats of Christmas and holiday fairs – to those signifying an everyday necessity in the 19th century as treacle in the workhouses and jam on factory hands' bread.

It is at this point that the nub of Mintz' argument lies. He insists that the practical precedes the symbolic, or as he puts it, that meaning is the *consequence* of activity. The messages encoded in sweetness could change as a result of increased usage. And the reason such increased usage, or activity, was possible is intricately bound up with the political interests, colonial aspirations and mercantile doings of nation states and their governments. 'Power' in the story of sugar links production and consumption, and operates, says Mintz at two levels. One is the point of use, the activity, that carries with it the power of the symbolic. The other is at the geopolitical level that made the use possible – a level at which, Mintz guesses, the English commoner was quite unaware. And it is at this higher level that the coincident push and pull of production and consumption are evident.

It is in this fashion that Mintz brings together the insights of structural anthropologists – approvingly recalling Lévi-Strauss' dictum that food is good to think – with a theorised view of the material realm. He grants that meaning and the symbolic is to be deciphered. But he refuses to leave it at that. Simply decoding the magnificent displays of sugar confections of the 17th century wealthy as signifying wealth stops the story before its end. Thus it forgets not only the 19th century poor's heavy dependence on the product, but risks losing sight of modern usages including the notorious sweetening of canned beans and meats of which consumers may often be only half aware. Mintz insists that an adequate anthropology – of food or anything else – cannot lose sight of history. He agrees that decoding rituals of eating and of the messages enshrined in food choice is vital; but what is needed is to

decode(s) the process of codification, and not merely the code itself.

(Mintz [82], p. 14.)

We can see, then, the parallels between Harris' and Mintz' perspectives. Mintz, however, takes a broader, eclectic and more complex view of the task. Where Harris focuses on a sort of cost-benefit analysis as practised by cultures, Mintz insists the analysis must include the social organisation of production, related patterns of consumption as well as all aspects of power. And where Harris merely dips into history to find apt illustrations and precursors to support his case, Mintz demands that an understanding of historical changes themselves be analysed. He offers into the bargain, an expanded style of anthropology that does not compare cultures on a static, geographically distant basis, but looks at their relationships as parts of a system ever more readily described in world terms.

Above all, Mintz again differs from Harris in his treatment of the symbolic sphere in eating habits. Certainly he is as clearly critical as Harris of structural approaches for putting symbols first, and he too turns the order round to give priority to the material. But where Harris states his case and then says little more about the relation between the symbolic and material realms, Mintz seeks to go further by incorporating attention to that very relation right through his analysis.

Two final words on this brand of materialist response must be added. Even if one is sympathetic to a Marxist-inspired analysis (and it is undoubtedly one means of bringing historical change, material social and economic forces as well as power simultaneously into view), Mintz' very well-written and careful analysis still leaves some nagging doubts. Ironically, it is the very fact that he is careful which gives rise to those doubts. He appears punctilious in repeatedly indicating where in the absence of appropriate evidence, he has to rely on guesswork; and more than once he points to the need for more anthropological fieldwork. But when he turns his attention to the ubiquitous place of sugar in modern American food habits, he arouses the suspicion that he is generalising overmuch from his own experience and point of view.

This worry surfaces towards the end of the book which concludes with a very elegant portrayal that links today's eating patterns with daily timetables. On the one hand he points to the astonishing versatility of sugar that can percolate all sorts of nooks and crannies of eating patterns, cuisines, and manufacturer's inventions. On the other he notes that contemporary experiences include the familiar feeling of always being rushed – and it is this feeling, he proposes, that makes us especially susceptible to the marketing efforts of what he faintly disparagingly refers to as the sucrose sellers.

This is all very recognisable, but is it quite the whole story? Despite cliché-ridden talk about the pace of modern life and the rhetoric of time- and labour-saving, feeling pressed for time is more variable in each direction for the whole population than for the professional classes to which he – and we – belong.

As he says, the official workday has not changed much in the last hundred years. But the official working day and an actual working day is not the same. First, remembering forecasts of the 'problem of leisure', the economy looks like no respecter of official accounting of working hours; currently continued uncertainty of employment is gloomily predicted for many industrialised countries. Second, gender cannot be overlooked: much of all unpaid work is done by women – an oversight that cannot be risked in the context of a discussion of food habits, the sphere where even though increased commercial manufacture has taken over much of the work (cf. Murcott [84]) a very great deal of activity indeed remains domestic and does not register in formal reckonings of workday length. (Here the sociological literature is burgeoning; for a brief introduction to part of it, see Murcott [87].) And third, there are variations over a lifetime: the young and the elderly probably experience timetables very differently from the woman fully employed running a home and caring for three pre-school age children, or from the construction worker filling 7 days a week with overtime.

These strictures are perhaps a little unfair. After all, Mintz continually presses the need for more research. But that in turn brings us up against another difficulty. As we have seen, he introduces a dynamic rather than static analytic angle to the study of food habits, and he seeks to incorporate the significance of the symbolic into his style of materialism. But he remains silent on the exceedingly difficult question of the social and cultural mechanisms whereby the symbolic arises. Thus the future fieldwork, that undoubtedly needs to be undertaken, is given little practical clue as to the most suitable framework to be adopted, or more particularly quite how we might research the way the symbolic is consequent upon activity.

The point remains, though, that despite such worries, the book is a major contribution to the discussion, serving to identify not only more precisely what needs researching but, as significantly, how we might develop adequately theorised viewpoints in so doing.

Goody

Turning to the third instance of a materialist response we come now to the work of the British social anthropologist Jack Goody, one of the two authors who considers whole cuisines rather than concentrating on the question of preferences, prescriptions and avoidances in the choice of selected food items. Although his substantive focus in this book, *Cooking, Cuisine and Class,* is thus different, Goody shares Mintz' conviction that the dynamic of history must be incorporated into anthropological perspectives on food habits. Moreover, he stoutly claims that an approach which takes account of comparisons between cultures is essential – indeed he views social anthropology as a branch of comparative sociology (a position indicated by his book's subtitle). Since all of human society and history is thus incorporated into the study of food habits, Goody's book is very wide-ranging indeed. It is also extremely densely written, not an easy read.

Goody is as critical of structuralism as Mintz and Harris (providing possibly the best exposition of its limitations of all four of its critics considered here). And he is aligned with them in not dismissing the importance of its insights – his sensitivity to the significance of meanings and codes of eating can be in no doubt after reading the account in this book (pp. 83–84) of his own experiences of coping with food shortages as a prisoner of war. Instead, he himself is interested in developing a perspective that *complements* a structural approach, filling in what it leaves out. In this respect his position contrasts with theirs: where they argue that the material is prior to the symbolic, Goody prefers, in general terms at least, to give primacy to neither.

For Goody, the type of explanation for culinary aspects of food habits is to be developed not so much in the structural approaches' 'languages' of the transformations of food, but by locating food and related practices in the context of the wider, and more diversified social structural arrangements of household and of class:

> (S)ince (food) is linked to the mode of production of material goods, the analysis of cooking has to be related to the distribution of power and authority in the economic sphere, that is, to the system of class or stratification and to its political ramifications.
>
> (Goody [51], p. 37.)

He specifies what he has in mind by summarising, as follows, the main areas of activity that cover the processes of providing and transforming food [51, p. 37]:

Processes	Phases	Locus
Growing	production	farm
Allocating/storing	distribution	granary/market
Cooking	preparation	kitchen
Eating	consumption	table
Clearing up	disposal	scullery

This scheme is further elaborated later in the book, allowing Goody to bring a gamut of considerations together – the types of foodstuffs available in any particular geographic region, the economic and political organisation that links who shall labour to produce food with who will partake of how much of it, when and with whom (both at a society-wide and domestic level), the technologies of both field and kitchen and so on.

Having set up the framework, we can now turn to consider the main theme of the book. This can be expressed simply enough, *Cooking, Cuisine and Class* is devoted to a question: Why does a 'high' (elite) and correspondingly 'low' (peasant) cuisine emerge in some societies but not others? It cannot merely be a matter of degrees of social and political complexity. For no high cuisine developed in most African 'traditional' societies, yet some of those societies are notable for their greatly differentiated social and political structures. Goody's strategy for tackling this question makes this book even more wide-ranging geographically and historically than Mintz'. Its sense of a world history and economy is as great, though perhaps less prominently the *subject* of the analysis.

He begins with a comparison of two North Ghanaian tribes, the LoDagaa and the Gonja, amongst whom he has extensive fieldwork experience. The many differences between the two are quite marked. For instance, the former had no tradition of chiefship, the latter a long history as an independent kingdom composed of a ruling estate, commoners as well as slaves. The LoDagaa are situated to the north of the Gonja, with different environmental conditions and thus agriculturally they differ. Rules of ceremony, hospitality and exchange are also different, in that the Gonja mark festive occasions by preparing and distributing food that is cooked, whereas at LoDagaa festivals it is distributed raw. For the LoDagaa, the basic diet – with little day-to-day variation – is a porridge of guinea-corn or millet, accompanied by a soup made mostly from ground-nuts or one or other type of leaf. Whereas in Gonja, yams form the basis for part of the year, and grains and cassava for the remainder, with more evidence of fish and wild meat as a relish.

And yet, despite such considerable differences of social structure in general, and in food in particular,

> the actual shape of the cuisine in both societies was surprisingly similar. We did not discover any major differantiation of cooking either on a periodic basis, for festivals, or in relation to the different strata of Gonja society. (Goody [51], p. 96).

At this point, Goody turns to those societies for which differentiation in cuisine is a hallmark. In a breathtaking sweep through Eurasia, he considers the evidence of Ancient Egypt and India, of ten centuries of China, right through to the development of industrialised eating in Europe and trends towards a world cuisine. He finally comes full circle, noting the impact of the world system in terms, among other things, of the demand for tinned sardines, tomato paste and bottled beer in the West Africa of his starting point. And all the while, he keeps the elements of his analytic scheme firmly in view. Goody turns up a number of answers to his book's original question. One of these arises out of taking a close look at the types of social and political differentiation within societies. Some of the African societies may be differentiated, but the differences are not that extreme: there are still some close links between the different strata, and sub-cultural differences are not so great. These characteristics lead Goody to call them 'hieratic' in contrast to differentiation of stratification systems such as Indian castes, mediaeval feudal estates or modern social class structures, which he describes as hierarchical.

Another answer takes account of differences in socio-cultural organisation related to variations in types of agriculture and the use of different implements. Thus Goody looks to Ethiopia, which may be geographically in Africa, but culturally lies outside. Here he can point to a conjunction of the development of hierarchical social and political organisation, the widespread use of the plough, *and* the emergence of a differentiated cuisine.

Two further answers are found in themes to which Goody [48, 50] has turned his attention in a number of his other books: the place of literacy and the social organisation and development of families and domestic groups [47, 49, 52]. Taking literacy first – in his account of the high and low of both cooking and hierarchical social arrangements, Goody draws on a variety of documentary sources, dietary and household manuals, treatises on health, as well as collections of recipes and cookbooks, of different eras in various societies. He treats them not only as sources of evidence about cuisines but, also, in their availability and use, as evidence about the societies of which they were/are a part. His point is not just that literacy is more closely associated with hierarchical societies whereas an oral tradition used to dominate hieratic societies, but that literacy can have multiple social consequences. It can be, as Goody puts it, an instrument of enlightenment – for instance it offers the potential for the complex elaboration of repertoires, including the culinary, and it can be a means whereby those aspiring to higher status can teach themselves the menus and manners of elite groups. But he also points out it can also be an instrument of oppression – until the advent of modern near-mass literacy, societies with writing were stratified by their access to the written word. Such *restricted* literacy serves to sharpen differences in life-style and taste in all manner of things, including, of course, cuisine.

Turning to the second theme, Goody shows that the organisation of households and the structures of kinship are qualitatively different between Eurasian and African societies. On this point once again his discussion is wide-ranging, covering not only the much remarked complex of correspondences and overtones between food and sex, not only the ties of obligation to provide for one another that link wives and husbands to each other and to their wider kin, but also in the conventions prescribing which (food-) work is done by whom for whom – i.e. the sexual division of labour. A key point here is the way in which for most African societies, even for those of high rank, kitchen work is done by wives to provide the homely dishes of domestic cuisine common to most households throughout the society. By contrast, food preparation for the elites and aristocracies of hierarchical societies is in the hands of servants and (male) professionals – consider modern social arrangements that on the whole casts men as chefs in the commercial sphere, women as cooks in the home.

Cooking, Cuisine and Class is an absorbing and important book. It is also demanding – despite Goody's attempt to write with a general rather than specialised readership in mind. And it defies brief summary. The work is massively condensed while theoretical and analytic considerations are continually woven into presentation of the data.

Finally, Goody's book may still seem an attractive proposition to those who are both frustrated by the limits of structuralisms and less sympathetic to Marxist approaches as an answer to the way we might tackle the study of food habits. Nonetheless, they will also find that the very breadth of Goody's work on cuisines loses (probably unavoidably) in depth and detail. For instance, he assumes too great a homogeneity of world/industrialised eating. And despite his proper insistence on accounting for historical change, the analysis retains an unexpectedly static feel. The jigsaw of pieces – taxes, tithes, agricultural implements and cooking utensils, surpluses and shortages, as well as the apparently more abstract but real (and constraining) enough features of divisions of labour, household and class, etc. – are assembled in a masterly fashion. But a view of how these pieces make up the picture, the dynamics of their articulation in actual practice, is not always clear – and left in the language of 'linked to' or 'related to'.

Having said all that, however, I must make one last, brief observation. *Cooking, Cuisine and Class* began its life as the idea for an inaugural lecture. It may very well turn out not to be Goody's last word on the matter.

Mennell

As noted earlier, Stephen Mennell is the only author, among the half-dozen whose work is considered in detail in this review, whose training and career are located within sociology rather than social anthropology. His response to structural approaches is far and away more dismissive than any of the three social anthropologists so far discussed and may well have something to do with this difference in background. Study and analysis of the symbolic in human affairs has long been a standing item on social anthropology's agenda. And there are different types of theoretical approach to it within the anthropological tradition – structural approaches being just one variety (cf. Firth [37] for a different stance on symbols and food and cf. also Appadurat [2], Hugh-Jones [56] and Laderman [61]). For sociologists, however, the topic is rarely the centre of close analytic attention. This is not to say that the significance of symbols is denied – far from it. Simply that the matter is assumed rather than explored in analytic detail. This – to anticipate his story for a moment – is just what Mennell does in *All Manners of Food* [78], when he talks, for instance, of the way the nobility of 17th century France indulged in lavish display at the table as one of the range of means used to distinguish themselves from those of lower social rank.

Mennell, however, is impatient with structuralisms for other reasons in addition to what, in his view, is a highly improbable mode of analysing symbols. Not only does he concur with all the assessments of wider limitations of the approaches noted at the end of an earlier section (p. 14; the omission of a historical perspective in particular) but he is also critical of an anthropological tendency to exaggerate the homogeneity of cultures that, for instance, plays down distinctions between social strata. For him, these limitations hinder the development of a style of explanation for food habits that takes adequate account of the evidence.

Consider two points: on the one hand we know that habits and taste in food changes (look no further than the last couple of decades in Britain where previously unknown versions of fast food have successfully found a place, never mind the massive differences over the last four or five centuries). On the other, a conservatism in food preferences is commonly noted (the British are stereotypically notorious in this respect). Existing explanations for such conservatism do not help us at all, says Mennell, since they amount to no more than circular statements that people eat what they prefer because they are accustomed to eating what they like.

Explanations for food choice have to proceed from a different starting point. The question for Mennell is not 'why do people display the preferences they do', but 'how do they have the chance to *become* used to what they eat?' In order to explore this question, Mennell devotes the book to an extended case study comparing the development since the middle ages of two closely related cuisines, the French and the English (he is careful to specify that he means English and not British, setting aside the Welsh, Scottish and Irish).

In reformulating the way we are to think about explanations of food choice, Mennell introduces his version of attention to the material social realm. Asking how people come to be able to express one or other set of preferences turns the spotlight on the complex amalgam of social and natural constraints and possibilities that shape people's lives at any given historical period. To explore this realm, Mennell turns not to Marxist nor anthropological traditions, but to one established by Norbert Elias, a European émigré to Britain.

In particular, Mennell follows the lead Elias [33, 34] sets in his major work *The Civilizing Process.* This work is devoted to an analysis of the history of the cultural shaping of psychic make-up. It argues that the extravagant and volatile mediaeval character could settle into more recognisable modern management of emotion that is self-controlled and discriminating, by virtue of the pacification of society (the reduction and control of the violence of wars and of the brutalities and gross uncertainties of day-to-day living) afforded by the transition from the feudal to the modern state. Elias' approach and method is foreign in many senses to English language sociologies (cf. Bauman [12]); but any further exposition here lies outside the scope of this review (for an introduction to his thought see Mennell [80]). What Mennell takes from Elias' work is a theory of how material social forces shape the expressions of emotion as manifest in life-style, manners and taste, and he applies this to account for changing food preferences and emerging cuisines.

Crudely put, his argument runs as follows: taste in eating, even appetite itself, is formed in the same way that Elias details the production of psychic make-up more generally. The transition from the feudal to the modern state shapes the parallel transition from the mediaeval extremes of feasts and famine to the emergence of emphasis on discrimination at the table. As the middle ages waned, food supplies began to improve. But more than that, the extension of trade, progressive division of labour and the process of state formation and internal pacification improved the *security* of those supplies.

Continued improvement meant that more sections of society were in a position to emulate an elite that hitherto was distinguished by, amongst other things, their 'gargantuan appetite'. Once food supply is more widely plentiful, the aristocratic 'blow out' stops: the nobility in both England and France of the 16th and 17th centuries could literally eat no more. This form of preserving distance between themselves and upstarts immediately beneath them in the social hierarchy reached its physical limit. The complex competition of shifting class structures, the development of greater internalised self-control rather than the brute capacity to stuff, is of a piece with the elevation of discrimination in eating to a socially valued position. Groaning boards became vulgar, ever more elaborate concoctions the height of good taste.

Thereafter shifts in what counts as good taste oscillate between pressures to the more and more fanciful that then collapse in being judged absurd as the virtues of simplicity are discriminatingly reasserted. But each cycle is not merely a resurrection of an earlier one: each has it own historical character, intricately made possible by social, economic and political changes. And by electing to compare England and France, Mennell helps us see more clearly the impact of those changes on shaping a cuisine.

A striking instance is his analysis of the differing relation the pre-revolutionary French aristocracy – with their preference for elaborate sauces and made dishes – had to real political power compared with their English counterparts – inclined to more simply dressed meats and fond of puddings. In France, 17th and 18th century aristocratic life centred ever more on the royal court; the nobility became more and more like the king's personal servants, less and less engaged in governmental and military power such that their lives were increasingly centred on elaborate social display and virtuoso consumption. By contrast, the English nobility had greater real political power and thus less preoccupation with

> outward symbols of differences in rank ... (less) pressure towards 'ornamental' display in matters such as manners, art, architecture and food. (Mennell [78], p. 121.)

Not only does this analysis help explain cross-Channel elite differences at that time, but the fate of the French aristocracy also shapes subsequent developments in French eating habits as well as the continued shaping of contrasts with the English. Along with these cycles in conceptions of good food, went shifts such that at times the tastes of the upper echelons of each country were closer to one another's than to their respective masses, while at others national differences become more pronounced. Then there is a complex interpenetration of French and English cooking, mutual admiration cross-cutting mutual antagonism. Above all, there is an overarching trend towards lessening differentiation between the food preferences of the social strata at the same time as the growth of greater width of choice for the whole populace – Elias' 'diminishing contrasts and increasing variety' – that brings us up to today.

This massively simplified summary does little justice to Mennell's argument. Nor does it convey the richness of his sources that range from catering trade journals, contemporary commentaries from different eras, cookbooks, historical studies of famine and of health, diaries, to women's magazines. Neither does it indicate the flashes of skill with which he marshalls those data demolishing inadequate conventional wisdoms as he goes – restaurants developed in France *before* the revolution. Or again, the idea that mediaeval cookery used lots of spices to hide the taint of meat that was off represents a modern blindness of judgement that disregards the evidence that surviving recipes of the period do not recommend fewer spices if the meat were fresh, and that in any case, the more selective use of flavourings developed long before effective improvements in techniques of meat preservation were invented.

Though very intriguing *All Manners of Food* is, in a manner reminiscent of Goody's, not an easy book. It has to be read attentively. As in Elias' work, Mennell cannot give the reader much guide to what is to come, for the method involves considerable familiarity with the material while simultaneously following the author's theorising. And the amount of that material – over which Mennell has very considerable mastery – is vast. Moreover, Mennell is so 'inside' that material that he gives pointers to his argument at a very abstract level. He does not always tie up a chapter in the light of the question he posed at its beginning and is inclined to introduce key bits of his argument in the middle of a paragraph – for example his theory of how gastronomy emerged as a bourgeois rather than courtly preoccupation. What is more, grasping what he has to say depends on a thorough appreciation of Elias' deceptively simple but, for very many, unfamiliar thesis.

Moreover, Mennell's exposition shares some of the problems already noted in Mintz' and Goody's work. First of all, as his account nears the present day, he comes close to failing to follow his (and Elias') precepts that changes in food preferences have to be analysed in relation to the range of prevailing social and political forces. Thus, although the early part of his case study is very convincing indeed, as he gets to later centuries, especially the 19th and 20th, he is less persuasive. It must be said, however, that later periods are a tougher prospect simply because of the sheer volume of material, much of which in this field has yet to receive a thorough scholarly going over.

So part of the problem is that Mennell is trying to deal with a study of primary sources at the same time as developing a much broader, secondary thesis. The result is that his examination of trade journals, women's magazines, cookbooks, etc. threatens to get taken out of context. For example, he bases his judgement that English school cookery in the 1890s was 'dull, stodgy and unimaginative (with) little skill demanded' [78, p. 231] on books of recipes used in the classroom. But would he not judge the whole curriculum of the period as dull and unimaginative? And he needs to remember that domestic science teaching, then and probably now, cannot be understood without reference to prevailing conceptions of proper roles for women, or to the belief that cookery was more appropriate for working-class than middle-class girls and in any case considered most suitable for the less able (cf. Dyhouse [32] and Purvis [96]).

Second, Mennell comes uncomfortably close to generalising from his own viewpoint. His treatment of domestic as opposed to professional cookery, especially in the present century, also proceeds without sufficient consideration of the 'power-politics' of marriages and households. Once again, gender is a more important consideration that he allows for in the array of forces shaping food preferences and expressions of taste (cf. Murcott [85, 87]). And it is hard not to suspect Mennell of some sort of partisanship in sounding as if he values traditions of French cuisine above the British. In his defence, though, it is probably fair to say that at the back of this is the view that would like to see everyone having the chance to learn to enjoy the taste of *hautes cuisines,* rather than a criticism of those who know little of them for the philistinism – a democratic rather than elitist judgement.

These and other difficulties must, however, be set alongside the assessment that *All Manners of Food* represents a pioneering lead in the sociological study of food habits. Mennell [79] himself is continuing to develop the work, though it is still too early to say whether others will be inspired to follow. Not only does it put the topic firmly at the centre of academic sociological concern, it does so within a theoretical framework that, although marginal to English language sociological work to date, should nonetheless command serious analytic attention as to *how* food habits shall be studied.

Concluding Remark

In this review I have selected from the still somewhat incohate literature in social anthropology and sociology on human food habits and eating patterns to highlight a detectable new development. Food habits were often enough ignored in these disciplines, or if considered, inclined to remain on the sidelines. With the work of the authors considered here, a serious debate over their nature, origins and above all study has begun. In particular, this debate explicitly focuses on the manner in which those patterns and habits shall be subjected to scholarly analysis.

The significance of the debate's emergence is, in sum, threefold. First, it indicates social anthropology's continued concern with food and eating and the more surely founded beginnings of such concern in sociology. Second, it may signal the move of food and eating to a more central position in the range of each discipline's topics of interest. Third, and most important, it means that the strengths of discussions and debates about the merits of different modes of analysis already well under way in each discipline, are explicitly and more systematically being drawn into the specific sociological and social anthropological literature on human food habits, preferences and cuisines.

Acknowledgements

The author is indebted to Graham Burchell, Sara Delamont, Teresa Rees and most particularly, Phil Strong for their comments during the preparation of this paper, and to Toby Murcott who drew the diagrams and helped check the typescript. She is also grateful once again to the staff at the School of Home Economics and at the Arts and Social Studies libraries of University College Cardiff for their help in locating material.

References

1 Adams, P.: Health of the state (Praeger, New York 1982).

2 Appadurat, A.: Gastro-politics in Hindu South Asia. Am. Ethnol. *8:* 494–511 (1981).

3 Aronson, N.: Nutrition as a social problem: a case study of entrepreneurial strategy in science. Soc. Probl. *29:* 474–487 (1982).

4 Arnott, M.L. (ed.): Gastronomy: the anthropology of food and food habits (Mouton, Den Haag 1975).

5 Aron, J.-P.: The art of using leftovers: Paris 1850–1900; in Forster, Ranum, Food and drink in history: selections from the annales (Johns Hopkins University Press, Baltimore 1979).

6 Atkinson, P.: From honey to vinegar: Lévi-Strauss in Vermont; in Morley, Wallis, Culture and curing: anthropological perspectives on traditional medical beliefs and practice (Owen, London 1978).

7 Atkinson, P.: The symbolic significance of health foods; in Turner, Nutrition and lifestyles (Applied Science Publishers, London 1980).

8 Atkinson, P.: Eating virtue; in Murcott, The sociology of food and eating: essays on the sociological significance of food (Gower, Aldershot 1983).

9 Ball, R.; Lilly, J.R.: The menace of margarine: the rise and fall of a social problem. Soc. Probl. *29:* 488–498 (1982).

10 Barker, T.C.; McKenzie, J.C.; Yudkin, J. (eds): Our changing fare: two hundred years of British food habits (McGibbon & Kee, London 1966).

11 Barthes, R.: Toward a psychosociology of contemporary food consumption; in Forster, Ranum, Food and drink in history: selections from the annales (Johns Hopkins University Press, Baltimore 1979).

12 Bauman, Z.: The phenomenon of Norbert Elias, Sociology *13:* 117–125 (1979).

13 Be Jong, W.: The stigma of obesity. J. Hlth soc. Behav. *21:* 75–87 (1980).

14 Boscom, W.R.: Yoruba food. Africa *21:* 41–53 (1951).

15 Bryant, C.A.; Courtney, A.; Markesbery, B.A.; DeWalt, K.M.: The cultural feast: an introduction to food and society (West, St. Paul 1985).

16 Burnett, J.: Plenty and want: a social history of diet in England from 1815 to the present day; revised ed. (Scolar Press, London 1979).

17 Charles, N.; Kerr, M.: Feeding children – the gap between theory and practice. Paper presented to the Health Education Council (1983).

18 Charles, N.; Kerr, M.: Food for feminist thought. Sociol. Rev. *34:* 537–572 (1986).

19 Cosminsky, S.: Alimneto and Fresco: nutritional concepts and their implications for health care. Human Organisat. *32:* 203–207 (1977).

20 De Garine, I.: Food is not just something to eat. Ceres *4:* 46–51 (1971).

21 De Garine, I.: The socio-cultural aspects of nutrition. Ecol. Food Nutr. *1:* 143–163 (1972).

22 De Garine, I.: Approaches to the study of food and prestige in Savannah tribes – Massa and Mussey of northern Cameroon and Chad. Soc. Sci. Informat. *19:* 39–78 (1980).

23 Douglas, M.: Purity and danger: an analysis of concepts of pollution and taboo (Routledge & Kegan Paul, London 1966).

24 Douglas, M.: Deciphering a meal. Daedalus, J. Am. Acad. Arts Sci. *101:* 61–81 (1972).

25 Douglas, M.: Social and religious symbolism of the Lele; in Douglas, Implicit meanings (Routledge & Kegan Paul, London 1975).

26 Douglas, M.: Structures of gastronomy; in Douglas. The future and the past, annual report (Russell Sage Foundation, New York 1976–77).

27 Douglas, M.: 'Beans' means 'thinks'. Listener *8:* 292–293 (1977).

28 Douglas, M.: In the active voice (Routlegde & Kegan Paul, London 1982).

29 Douglas, M. (ed.): Food in the social order: studies of food and festivities in three American communities (Russell Sage Foundation, New York 1984).

30 Douglas, M.; Nicod, M.: Taking the biscuit: the structure of British meals. New Soc. *30:* 744–747 (1974).

31 Douglas, M.; Isherwood, B.: The world of goods (Penguin, Harmondsworth 1978).

32 Dyhouse, C.: Girls growing up in late Victorian and Edwardian England (Routledge & Kegan Paul, London 1981).

33 Elias, N.: The civilizing process: the history of manners (Blackwell, Oxford 1978); Über den Prozess der Zivilisation, vol. 1 (Falken-Verlag, Basel 1939).

34 Elias, N.: State formation and civilization (Blackwell, Oxford 1982); Über den Prozess der Zivilisation, vol. 2 (Falken-Verlag, Basel 1939).

35 Farb, P.; Armelagos, G.: Consuming passions: the anthropology of eating (Houghton Mifflin, Boston 1980).

36 Fenton, A.; Owen, T.M. (eds): Food in perspective (Donald, Edinburgh 1981).

37 Firth, R.: Food symbolism in a pre-industrial society; in Symbols, public and private (Allen & Unwin, London 1973).

38 Firth, R.: Housekeeping among Malay peasants; 2nd ed. (Athlone Press, London 1966); 1st ed. (1943).

39 Fischler, C.: Gastro-nomie et gastro-anomie: sagesse du corps et crise bioculturelle de l'alimentation moderne. Communications *31:* 189–210 (1979).

40 Fischler, C.: Food habits, social change and the nature/culture dilemma. Soc. Sci. Informat. *19:* 937–953 (1980).

41 Fischler, C.: The Socratic approach. The Guardian *January 13* (1980).

42 Fischler, C.: Food preferences, nutritional wisdom and sociocultural preferences; in Walcher, Kretchmer, Food nutrition and evolution (Masson, New York 1981).

43 Fitzgerald, F. (ed.): Nutrition and anthropology in action (Van Gorcum, Assen 1976).

44 Forster, R.; Ranum, O.: Food and drink in history: selections from the annales (Johns Hopkins University Press, Baltimore 1979).

45 Fortes, M.; Fortes, S.L.: Food in the domestic economy of the Tallensi. Africa *9:* 237–276 (1936).
46 Gelfand, M.: Diet and tradition in an African culture (Livingstone, Edinburgh 1971).
47 Goody, J.: The developmental cycle in domestic groups (Cambridge University Press, Cambridge 1958).
48 Goody, J.: Literacy in traditional societies (Cambridge University Press, Cambridge 1968).
49 Goody, J.: The character of kinship (Cambridge University Press, Cambridge 1974).
50 Goody, J.: The domestication of the savage mind (Cambridge University Press, Cambridge 1977).
51 Goody, J.: Cooking, cuisine and class: a study in comparative sociology (Cambridge University Press, Cambridge 1982).
52 Goody, J.: The development of the family and marriage in Europe (Cambridge University Press, Cambridge 1983).
53 Harris, M.: Cows, pigs, wars and witches: the riddles of culture (Fontana, Glasgow 1977).
54 Harris, M.: Cannibals and kings: the origins of cultures (Fontana, Glasgow 1978).
55 Harris, M.: Good to eat: riddles of food and culture (Allen & Unwin, London 1986).
56 Hugh-Jones, C.: Food for thought: patterns of production and consumption in Pira-Parana society; in La Fontaine, Sex and age as principles of social differentiation (Academic Press, London 1978).
57 James, A.: Confections, concoctions and conceptions; in Wates, Popular culture: past and present (Croom Helm, London 1981).
58 Jerome, N.W.; Kandel, R.F.; Pelto, G.H. (eds): Nutritional anthropology: contemporary approaches to diet and culture (Redgrave, New York 1980).
59 Johnson, J.P.: A hundred years eating: food, drink and the daily diet in Britain since the late nineteenth century (Gill & Macmillan, Dublin 1977).
60 Kerr, M.; Charles, N.: Food as an indicator of social relations. Proc. British Sociological Association Annu. Conf., Manchester 1982.
61 Laderman, C.: Symbolic and empirical reality: a new approach to the analysis of food avoidances. Am. Ethnol. *8:* 468–493 (1981).
62 Larkin, J.C.; Pines, H.A.: No fat persons need apply: experimental studies of the overweight stereotype and hiring preference. Sociol. Work Occupat. *6:* 312–327 (1979).
63 Leach, E.: Culture and communication: the logic by which symbols are connected (Cambridge University Press, Cambridge 1976).
64 Leach, E.: Lévi-Strauss (Fontana, London 1970).
65 Leitzmann, C.; Oltersdorf, U.: Introductory remarks and definitions; in Diehl, Leitzmann, Measurement and determinants of food habits and food preferences. EURO-NUT: a concerted action project on nutrition in the European Community, Report 7, Wageningen 1986.
66 Lévi-Strauss, C.: The culinary triangle, New Soc. *December:* 937–940 (1966); Le triangle culinaire. L'Arc *26:* 19–29 (1965).
67 Lévi-Strauss, C.: The raw and the cooked (Cape, London 1970); Le cru et le cuit. Mythologiques I (Librairie Plon, Paris 1964).

68 Lévi-Strauss, C.: From honey to ashes (Cape, London 1973); Du miel aux cendres. Mythologiques II (Librairie Plon, Paris 1966).
69 Lévi-Strauss, C.: The origin of table manners (Cape, London 1978); L'origine des manières de table. Mythologique III (Librairie Plon, Paris 1968).
70 MacAndrew, C.; Edgerton, R.B.: Drunken comportment: a social explanation (Nelson, London 1970).
71 Marett, R.R.: Food rites; in Evans-Pritchard, Firth, Malinowski, Schapera, essays presented to C.G. Seligman (Kegan Paul, Trench Trubner, London 1934).
72 Mass Observation: The pub and the people: a worktown study (Gollancz, London 1943).
73 Mead, M.: The factor of food habits. Annls Am. Acad. polit. soc. Sci. *225:* 136–141 (1943).
74 Mead, M.: Dietary patterns and food habits. J. Am. diet. Ass. *19:* 1–5 (1943).
75 Mead, M.: Cultural pattening of nutritionally relevant behavior. J. Am. diet. Ass. *25:* 677–680 (1949).
76 Mead, M.: We don't like what we don't eat. Cyprus med. J. *9:* 90–93 (1957).
77 Mead, M.: The changing significance of food. Am. Scient. *58:* 176–181 (1970).
78 Mennell, S.: All manners of food: eating and taste in England and France from the Middle Ages to the present (Blackwell, Oxford 1985).
79 Mennell, S.: Eating in the Netherlands. De Gids *January* (1987).
80 Mennell, S.: Norbert Elias and figurational sociology (Blackwell, Oxford, in press).
81 Mintz, S.W.: Time, sugar and sweetness. Marxist Perspect. *2:* 56–73 (1979).
82 Mintz, S.W.: Sweetness and power; the place of sugar in modern history (Viking, New York 1985).
83 Murcott, A.: On the social significance of the 'cooked dinner' in South Wales. Soc. Sci. Informat. *21:* 677–695 (1982).
84 Murcott, A.: Women's place: cookbooks' image of technique and technology in the British kitchen. Women's Stud. Int. Forum *6:* 33–39 (1983).
85 Murcott, A.: 'It's a pleasure to cook for him...': food, mealtimes and gender in some South Wales households; in Gamarnikow et al., The public and the private (Heinemann, London 1983).
86 Murcott, A. (ed.): The sociology of food and eating: essays on the sociological significance of food (Gower, Aldershot 1983).
87 Murcott, A.: Opening the 'black box': food, eating and household relationships. Sosiaalilaaketieteellinen Aikakauslehti. J. soc. Med. *23:* 85–92 (1986).
88 Murcott, A.: You are what you eat: exploration of anthropological issues influencing food choice; in Ritson, The food consumer (Wiley, Chichester 1986).
89 Nichter, M.: Drink boiled water: a cultural analysis of a health education message. Soc. Sci. Med. *21:* 667–669 (1985).
90 Nicod, M.: A method of eliciting the social meaning of food; MPhil. thesis, London (1974).
91 Nicod, M.: Food studied as a medium of communication; in Turner, Nutrition and lifestyles (Applied Science Publishers, London 1980).
92 O'Lauchlin, B.: Mediation of contradiction: why Mbum women do not eat chicken; in Rosaldo, Lamphere, Women, culture and society (Stanford University Press, Stanford 1974).

93 Oddy, D.J.; Miller, D.S.: The making of the modern British diet (Croom Helm, London 1976).
94 Oddy, D.J.; Miller, D.S. (eds): Diet and health in modern Britain (Croom Helm, London 1985).
95 Pelto, G.H.: Current research directions in nutritional anthropology. Sosiaalilaaketieteellinen Aikakauslehti J. soc. Med. *23:* 93–103 (1986).
96 Purvis, J.: Domestic subjects since 1870; in Goodson, Social histories of the secondary curriculum: subjects for study (Falmer Press, Lewes 1985).
97 Richards, A.I.: Hunger and work in a savage tribe: a functional study of nutrition among the southern Bantu (Routledge, London 1932).
98 Richards, A.I.: Land, labour and diet in northern Rhodesia (Oxford University Press, London 1939).
99 Robertson Smith, W.: Lectures on the religion of the Semites (Appleton, New York 1889).
100 Robinson, D.: From drinking to acoholism: a sociological commentary (Wiley, London 1976).
101 Robson, J.R.K. (ed.): Food, ecology and culture: readings in the anthropology of dietary practices (Gordon & Breach, New York 1980).
102 Roth, J.A.: Health purifiers and their enemies: a study of the Natural Health movement in the United States with a comparison to its counterpart in Germany (Prodist, New York 1976).
103 Rowntree, B.S.: Poverty: a study of town life (Macmillan, London 1901).
104 Rowntree, B.S.: The human needs of labour; revised ed. (Longmans Green, London 1937).
105 Sahlins, M.: Stone age economics (Tavistock, London 1974).
106 Sanderson, M.E.; Winkler, J.T.: Strategies for implementing NACNE recommendations. Lancet *8363:* 1353–1354 (1983).
107 Simons, F.J.: Eat not this flesh: food avoidances in the Old World (University of Wisconsin Press, Wisconsin 1961).
108 Townsend, P.: Poverty in the United Kingdom: a survey of household resources and standards of living (Penguin, Harmondsworth 1979).
109 Troyer, R.J.; Markle, G.E.: Coffee drinking: an emerging social problem? Soc. Probl. *31:* 403–416 (1984).
110 Turner, B.S.: The discourse of diet. Theor. Cult. Soc. *1:* 23–33 (1982).
111 Turner, B.S.: The government of the body: medical regimens and the rationalization of diet. Br. J. Sociol. *33:* 254–269 (1982).
112 Twigg, J.: Food for thought: purity and vegetarianism. Religion *9:* 13–35 (1979).
113 Wellin, E.: Child feeding and food ideology in a Peruvian village; mimeo (WHO, Genève 1953).
114 Wellin, E.: Maternal and infant feeding practices in a Peruvian village. J. Am. diet. Ass. *31:* 889–894 (1955).
115 Wellin, E.: Water boiling in a Peruvian town; in Paul, Health, culture and community: case studies of public reactions to health programs (Russell Sage, New York 1955).
116 Wiese, H.J.C.: Food behavior in Haiti. Human Organisat. *35:* 193–200 (1976).
117 Wilson, M.: The wedding cakes: a study of ritual change; in La Fontaine, The interpretation of ritual: essays in honour of A.I. Richards (Tavistock, London 1972).

118 Winkler, J.; Sanderson, M.: Eating our way to health. New Soc. *65:* 33 (1983).
119 Worsley, P.: The utilization of natural food resources by an Australian Aboriginal tribe. Acta Ethnogr. *10:* 153–190 (1961).
120 Wuthnow, R.; Hunter, J.D.; Bergesen, A.; Kurzweil, E.: Cultural analysis: the work of Peter L. Berger, Mary Douglas, Michel Foucault and Jurgen Habermas (Routledge & Kegan Paul, London 1984).
121 Young, M.W.: Fighting with food (Cambridge University Press, Cambridge 1971).
122 Yudkin, J.; McKenzie, J.C. (eds).: Changing food habits (McGibbon & Kee, London 1964).

Anne Murcott, Department of Sociology, University College Cardiff,
PO Box 78, Cardiff CF1 1XL (UK)

Wld Rev. Nutr. Diet., vol. 55, pp. 41–57 (Karger, Basel 1988)

Feeding Practices in Colombia

Analysis of a National Household Survey (DANE, DNP PAN, 1981)

Maria Eugenia Romero [1]

Independent Researcher and Consultant, Apartado Aereo 39372,
Bogotá, D.E.1, Colombia, South America

Contents

I. Background

The study of feeding practices and patterns has recently begun in Colombia. Interdisciplinary research projects have been developing with the participation of nutritionists, statisticians and anthropologists.

The purpose of this article is to describe some of the feeding practices in Colombia as derived from the National Feeding and Housing Survey performed by the National Department of Statistics (DANE), and the National Department of Planning (DNP PAN) in 1981. The analysis that follows was developed in 1985 [1].

[1] Licenciado in Anthropology, Universidad de Los Andes, Bogotá (1970). Master of Arts, Anthropology, Southern Illinois University, Carbondale, Ill. (1975).

Several national feeding surveys have been done in the country but only until 1981. A National Survey representing 98.7% of the total population was shown to be representative of five major geographical regions of the country both in the rural and urban areas. Some other qualitative studies in the field of feeding habits and practices have been reviewed concluding that many of them were never finished or published [2]. There is a lack of information concerning traditional peasant and indigenous feeding practices and their nutritional values. Some recent anthropologically oriented case studies identified traditional peasant foods and recipes with adequate content of iron, vegetable protein and calories [3].

Some findings in the rural areas of the country stress the relationship between annual food production, social and cultural factors and consumer practices. High levels of malnutrition, infant mortality and morbidity still occur, both in marginal rural areas and in urban slums, as a consequence of poor living conditions.

However, many factors, including income, food availability and rather easy access to processed foods, are inducing rapid changes, not only in infant feeding but also in the household diet. Prestige associated with particular types of foods and the influence of the mass media (i.e. radio and television) also play a role in changing food preferences and consumer behavior. Increasingly, it is only older people, both women and men, who have knowledge about the traditional culture of food.

Changes are also taking place in the way foods are purchased. Shops and small grocery stores in low and middle income neighborhoods usually sell on credit, which means that families limit their diets to what is available in these stores. The only solution for migrant families is to adapt their eating to their new economic and environmental conditions, and, hence, accept new products. 'People eat more frequently what is available to them, and, unfortunately, if the cheapest foods are those without nutritional quality these are selected and consumed with consequent effects on the nutritional status of the population' [4].

There seems to be similar levels of under-consumption (about 40%) both in rural and in urban areas [5]. 'There are very serious nutritional problems in Colombia. Among those we could mention protein-calorie malnutrition, chronic adult malnutrition and iron deficiencies' [6]. Protein-calorie malnutrition in the population under 5 years of age reaches 65% [7].

The food industry is the single largest industry in Colombia, with a GNP of 37%. In 1979, a total of 115,821 tons of food were imported, a figure that continues to the present. The food industry grew by only 5% between 1970 and 1975. Thus, it shows signs of stagnation, and the only solution is to produce more and better foods and to adapt production to income.

The overview for the agricultural sector is not more promising: a growth rate of 4% in 1975–1979 as compared to 5% in 1970–1975 [8]. The relationship between the agricultural sector and industry becomes more difficult each day because there has not been policy coordination. Recent statistics are even more dramatic: there has been a decrease of 8% in cultivation area in the first semester of 1984 compared to the same period in 1983, as judged from loan data from the National Agriculture Institute. The crops with highest reductions in cultivation are rice, barley, beans, sorgum, soybeans and vegetables. During the first part of 1984 food prices increased more rapidly than the average rate of increase from previous years [9].

Some recent data demonstrate the crisis of the agriculture and cattle raising sector. High cost of fertilizers and scarcity of animal foods, added to frosts and prolonged dry and rainy seasons, have created serious conditions for the production of main staples. In 1984–1985 the cost of fertilizers increased by 35% [10]. Both rice, soybeans and sorgum demonstrated very low levels of crop area [11].

The worldwide concern about food production and availability has focussed on problems related to food scarcity, considered a world crisis similar to the energy crisis. Worldwide distribution of main staples and foods aggravates this condition. Colombia, a country of 1,138,490 km^2 and with an estimated population of 33 million in 1985, is not unaware of this crisis: the situation should be explained taking into account political, economic and sociocultural aspects.

Anthropology has contributed much to the understanding of food problems and issues. Several topics related to human nutritional needs and stability of food habits, among others, may be analyzed when a convergence of methods and theories of anthropology and economics, for example, is seeked [12].

The Colombian population settled traditionally in the Andean area and developed agriculture on the slopes of the hills. Rural areas and urban centers are more densely populated in the Inter-Andean valleys of the country's three mountain ranges. Before 1940, 70% of the population lived in the rural areas; by 1979 only 40% remained in the rural areas, while 60% lived in the cities.

The relatively low level of industrial development began increasing after 1930 when foreign investment was promoted in the country. Cities such as Medellín, Cali, Barranquilla and Bogotá became industrial centers, attracting people from the rural areas. Industrial development was tied to changes in land ownership relations in rural areas. Changing economic conditions played an important role in the participation of women in the labor force, both in the formal and informal sectors of economy. These changes led to further transformation in social and family relations.

II. The National Housing and Feeding Survey (1981)

This survey estimated a universe of 25,875,343 inhabitants, 65.2% of which lived in urban areas and 34.8% in rural areas. The country was divided into five major geographical regions excluding the Orinoco lowlands and Amazon basin. Each region had a minimum of three million inhabitants: the total sample of representative households was 10,000 with a subsample on dietary practices the day before the survey was performed. The major regions were Atlantic Eastern, Bogotá, Central and Pacific (fig. 1).

Fig. 1. Regions and departments in Colombia

Region	Department	Region	Department
Atlantic	Atlántico	Bogota D.E.	Antioquia
	Bolivar	Paisa	Caldas
	César		Quindio
	Córdoba		Risaralda
	La Guajira		
	Magdalena	Central	Caquetá
	Sucre		Huila
			Tolima
Eastern	Boyacá		
	Cundinamarca	Pacific	Cauca
	Meta		Chocó
	Norte de Santander		Nariño
	Santander		Valle

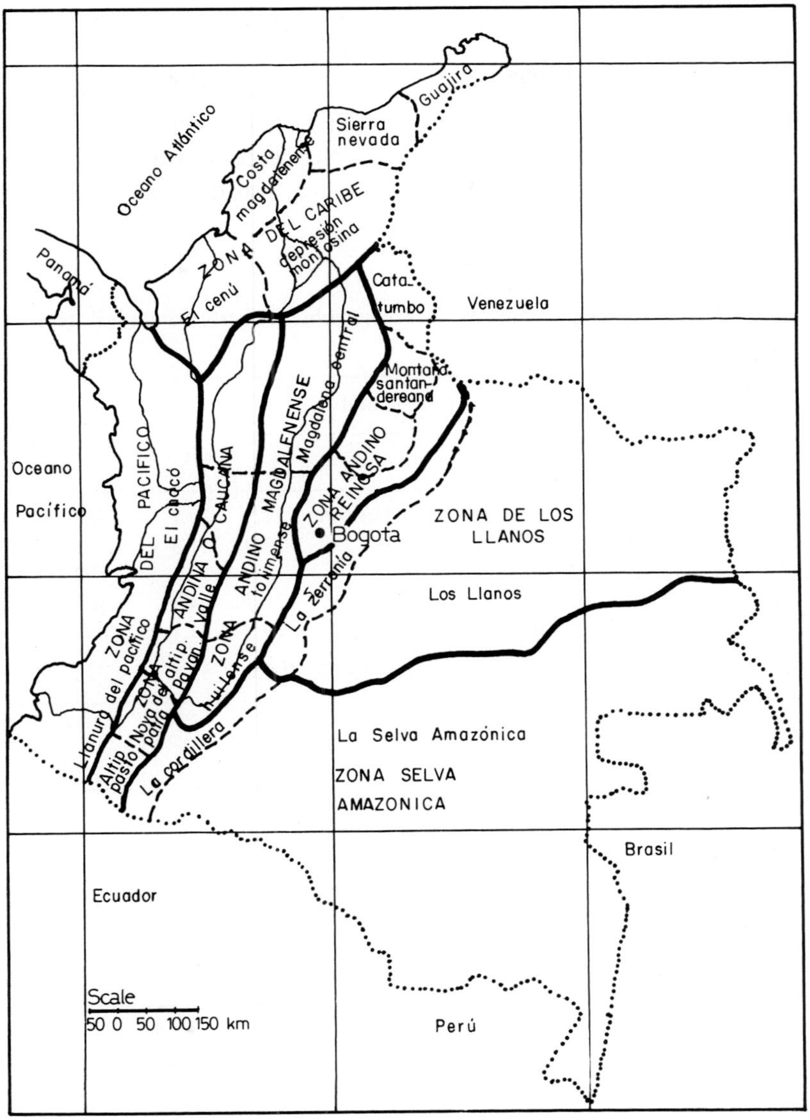

Fig. 1. Regions and departments in Colombia (for explanation see p. 44).

III. Characteristics of the Population

According to the survey, 48.7% of the population are males and 51.3% females; 39.4% of this population is under 15 years of age and 54.6% of the total population is economically active. Six percent of the population are aged 60 years or more. By the time the survey was developed 59.7% of the

Table I. Geographical regions and number of households in urban and rural areas: Colombia, 1981

Region	Urban			Rural			Total	
		%[1]	%[2]		%[1]	%[2]		%[1]
Atlantic	523,088	16.3	60.9	336,406	21.5	39.1	859,494	18.0
Eastern	444,368	13.9	48.7	467,389	29.9	51.3	911,757	19.1
Bogotá	802,866	25.0	100	–	–	–	802,866	16.8
Central	859,466	26.8	65.6	449,711	28.7	34.4	1,309,177	27.4
Pacific	577,016	18.0	64.9	311,921	19.9	35.1	888,937	18.6
Total	206,804	100.0	67.2	1,565,427	100.0	32.8	4,772,231	100.0

[1] Percentage per column.
[2] Percentage per aisle.

Table II. Household distribution and levels of monthly income in urban and rural areas: Colombia, 1981

Monthly income in Colombian pesos[1]	Urban, %	Rural, %	Total, %
Up to 5,310	13.3	31.8	19.3
5,311–10,000	25.2	28.1	26.2
10,001–15,000	15.3	12.0	14.2
15,001–20,000	9.1	5.4	7.9
20,001–40,000	16.0	6.0	12.7
40,001 or more	7.5	1.4	5.5
Without income or no information	13.6	15.3	14.2
Total	100.0	100.0	100.0

[1] In 1981 the currency rate was $53.08 to a dollar (May).

households had a monthly salary under three minimum wages (in 1981 the minimum monthly wage was $7,200.00 Colombian pesos). In the rural areas this percentage rose to 71.9%.

Although the living conditions in the country are not the worst among the developing countries yet, some figures are worrying, as shown in tables I and II: 40.5% of the households do not own a house: 60% of the total number of households entirely lack sewage, water and electricity. Only 16% of the households in the country have adequate water supply and 11% have a sewage system [13].

Table III. Foods most commonly used (24 h recall performed in a subsample) in urban and rural areas: DANE DNP DRI PAN Survey, 1981

Food consumed	Percentage of households		
	urban	rural	total
Salt	88.8	93.4	90.3
Onion	83.6	80.4	82.6
Rice	83.0	76.1	80.7
Potato	76.1	72.0	74.8
Beef	72.0	63.2	69.0
Vegetable oil	72.8	48.6	64.9
Liquid milk	66.2	53.7	62.1
Plaintains	59.7	64.9	61.4
Panela	57.9	65.8	60.5
Tomato	67.5	35.9	57.2
Sugar	63.4	39.0	55.5
Chocolate	49.6	39.2	46.2
Bread	56.4	24.1	45.8
Eggs	46.9	32.9	42.3
Corn	41.2	43.6	41.9
Manioc	30.2	49.3	36.4
Carrots	29.5	13.3	24.2
Spaghetti	22.9	25.3	23.7
Beans	20.9	26.6	22.8
Soda pop	13.0	5.2	10.5
Yam	7.5	4.2	6.4
Oranges	7.0	3.2	5.8
Bananas	6.1	4.1	5.5

IV. Foods Most Commonly Used

The results of the survey show that in more than 50% of the households the foods consumed the day previous to the survey included rice, potatoes, beef (includes all sorts of meats and bones from cattle), vegetable oil, plaintains, *panela* (solid brown sugar), tomato, sugar, salt and onion (table III). Some foods are consumed more in the households of the urban areas, such as beef, vegetable oil, tomatoes, bread, eggs and soda pop. Among the foods most commonly used in the rural areas as compared to the urban we could mention plaintains, panela, manioc and beans.

Table IV. Calorie contribution of the foods most commonly used (%)

Type of food	Rural	Urban	Total
Rice	17.6	18.8	18.3
Plaintains	12.8	7.6	9.5
Potatoes	9.4	7.4	8.1
Corn	9.9	6.1	7.5
Vegetable oil	5.0	8.9	7.4
Beef	5.8	6.5	6.2
Bread	2.1	7.0	5.2
Sugar	3.3	6.0	5.0
Manioc	9.0	2.6	4.9
Liquid milk	4.0	4.4	4.3
Beans	2.4	2.3	2.3
Spaghetti	2.1	2.1	2.1
Chocolate	1.7	2.3	2.1
Vegetable lard	2.5	1.8	2.1
Eggs	1.0	1.9	1.6
Pork lard	0.9	1.2	1.1
Cheese	1.2	0.9	1.0
Wheat	1.2	0.6	0.8
Powder milk	0.2	1.0	0.7
Biscuits, cakes and cookies	0.4	0.8	0.7
Total	92.5	90.1	91.0

V. Percentual Contribution of Calories and Proteins

The nutrient intake is given by the amount of foods consumed and the nutritional value of the foods [14]. Rice contributes with 18.3% of the calorie intake; then follows plaintains and potatoes. When we compare the urban and the rural areas of the country we could say that the main source of calories in the rural households are plaintains, potatoes, corn and manioc, while in the urban households the main source of calories comes from rice, vegetable oil, bread, and sugar (table IV). Beef and rice contribute with 27.3% of protein in the total number of households. Vegetable protein is given by foods such as rice, corn, beans, peas and lentils (table V).

Table V. Protein contribution of the foods most commonly used (%): Colombia, 1981

Type of food	Rural	Urban	Total
Beef	26.4	27.8	27.3
Rice	13.4	13.4	13.4
Liquid milk	7.9	8.2	8.1
Potatoes	7.5	5.5	6.2
Corn	8.5	4.8	6.1
Beans	5.7	5.0	5.2
Bread	2.0	6.2	4.7
Eggs	2.7	4.9	4.2
Plaintains	4.1	2.2	2.9
Cheese	2.8	2.3	2.5
Spaghetti	2.3	2.1	2.2
Fish	3.5	1.5	2.2
Poultry	1.1	2.4	2.0
Powder milk	0.5	2.0	1.5
Peas	1.5	1.4	1.4
Wheat	1.8	0.8	1.1
Manioc	1.7	0.5	0.9
Pork	0.7	1.0	0.9
Lentils	0.3	1.0	0.7
Chocolate	0.6	0.7	0.6
Total	94.9	93.7	94.1

Table VI. Average consumption of calories and nutrients-levels of adequacy per region and area: Colombia, 1981

Region/area	Calories	Proteins	Calcium	Iron	Vitamin A	Thiamine	Riboflavin	Niacin	Vitamin C
Atlantic	106.0	113.3	50.2	96.4	57.5	71.0	85.6	84.6	239.2
Urban	106.9	116.4	50.8	98.9	63.7	74.3	92.8	88.3	229.9
Rural	104.4	108.5	49.1	92.4	47.9	65.9	74.5	78.7	252.9
Eastern	94.3	100.8	55.8	120.7	41.7	94.9	97.3	87.9	258.9
Urban	93.0	101.2	54.8	115.5	45.4	83.4	92.6	81.0	213.4
Rural	95.7	100.5	56.8	125.9	38.0	106.3	101.9	94.8	304.6
Bogota	92.0	104.0	59.3	112.7	61.2	81.8	100.0	79.8	215.5
Central	98.3	103.7	51.7	110.7	48.4	81.7	79.8	76.5	167.9
Urban	96.4	105.1	55.7	109.9	51.0	78.2	84.1	75.9	157.9
Rural	102.0	101.1	43.9	112.4	43.3	88.3	71.5	77.6	187.1
Pacific	101.6	100.9	44.8	101.7	61.8	82.2	80.3	85.0	229.5
Urban	105.0	109.3	51.7	105.0	65.5	79.4	87.0	88.2	209.6
Rural	95.9	86.6	33.1	96.0	55.4	86.9	68.9	79.5	264.1
Total	98.5	104.4	52.2	108.7	53.3	82.4	87.6	82.2	217.0
Urban	98.0	106.8	55.0	108.7	57.3	79.4	91.2	81.7	200.4
Rural	99.4	99.6	46.6	108.7	45.2	88.4	80.4	83.2	250.7

VI. Adequacy in the Levels of Consumption of Calories and Nutrients

The analysis of the levels of adequacy of calories and nutrients is one of the methods of identifying nutritional problems in a population. The adequacy is given by the amount of foods consumed at the household on the day previous to the survey and the recommendations about the daily allowances [15]. The RDA was determined by household according to the number of members, their age, sex and physiological status.

It is understood that the RDA are given for healthy individuals; an adequate consumption level of calories, proteins and nutrients leads to a healthy status.

The information analyzed below concerns the levels of adequacy in calories and nutrients according to regions and areas in the country.

Adequacy of Calories and Nutrients

The information shows that the average adequacy of calories is 98.5% for the total population, this adequacy is higher in the rural area. This adequacy is over 100% in the Atlantic and Pacific regions; the lowest level is found in Bogotá (92%) (table VI).

In respect to proteins, this adequacy is 104.4% for the total population and it is higher in the urban area. There are regions where an adequacy higher than 100% exists. In the rural area of the Pacific region this adequacy is under 100%, i.e. 85.6% (table VI).

The level of adequacy of calcium is 52.2%, with a lowest figure in the rural areas (46.5%). The region with the lowest level of adequacy is the Pacific with 44.8% with even a lowest figure in the rural area of this region.

The levels of adequacy of iron are 108.7%. The Atlantic region is the one where there is a figure under 100% and it has the lowest value in the rural areas of this region. Is it important to mention that the biological utilization of this mineral, as it comes from animal or vegetable foodstuffs, may be affected by intestinal parasites, a problem which was not identifiable in this study.

The level of adequacy of vitamin A is under 53.7% and it is lower in the rural area. Levels of adequacy under the national total are found in the Eastern and Central regions. These levels are always lower in the rural areas in respect to the urban areas. In respect to thiamine there is a total level of adequacy in the consumption at the national level of 82.4% and there is a lower figure in the urban area. The Atlantic region presents the lowest levels of adequacy of thiamine (71%).

The levels of adequacy in the consumption of riboflavin are 87.6% and these levels are lower in the rural areas (80.4%). The lowest levels are found in the Pacific and Central regions. Bogotá and the rural area of the Eastern region present levels of adequacy of this nutrient of 100% (table VI).

The levels of adequacy in the consumption of niacin are 82.2% for the total population although lower figures are found in the Central region and in Bogotá.

The levels of adequacy of Vitamin C are 217%. These levels are higher in the rural areas. It is important to mention that this analysis was performed with raw food and does not show the losses due to cooking.

VII. Main Preparations Consumed in the Household

The study of feeding practices and patterns gathered data according to the preparations used in the households on the day previous to the survey. Information was gathered about preparations consumed for breakfast, lunch and dinner. Information concerning other instances at which foods are consumed did not demonstrate statistical relevance (i.e. morning and afternoon snacks). Before the analysis was performed some preparations were grouped in order to attain a better understanding of the feedig practices. For example, the category corn tortilla (*arepas de maiz*) includes several types of *arepas* prepared all over the country. There may be 20 or more types of this preparation. Preparations made out of the home included bread, biscuits, cookies, soda pop and any other food not made at home. *Sancocho* is a stew made of manioc, plantains, potatoes, corn meat, or fish or chicken; this preparation differs according to area. The category *beef* included steak, roundsteak, ground meat, flanksteak, boiled meat, veal, eye of round, entrails, tongue, and bones. *Mazamorra* as a category includes heavy corn or plaintain soups, sweet or salty, salads included all kinds of vegetables, the category soups of grains and cereals included soups of barley, wheat and oats. Roots and plaintains included potato, manioc, *arracacha* (arrowroot) yam and several kind of plaintains also *batata* and *bore.*

The category rice with other foods included foodstuffs such as meat, vegetables, coconut, raisins, spaghetti, pork, chicken, kidney beans, plaintains, eggs, fish.

Rice is consumed in a high percentage of the households: 77% urban and 64.9% rural. Drinks such as chocolate and *agua de panela* are consumed in 74.9 and 52.2% of the households in the urban area and (table VII) 40.9 and 45.7% of the households of the rural area. Some preparations such as black coffee, plaintains, coffee with milk, potatoes, eggs, *arepas,* broths and juices are used in more than one-quarter of the households. Differences exist

Table VII. Main preparations consumed in the rural and urban areas: Colombia 1981

Preparations	Households		
	urban %	rural, %	total, %
Rice	77.0	64.9	73.0
Preparations made out of the home	74.9	37.2	62.4
Chocolate	52.2	40.9	48.5
Agua de panela	42.4	45.7	43.5
Beef	47.2	34.9	43.1
Black coffee	34.1	50.5	39.5
Plaintains	40.1	34.0	38.1
Coffee with milk	39.6	25.7	35.0
Eggs	40.8	21.9	34.6
Potato	33.1	25.6	30.6
Arepas	28.2	32.8	29.7
Broths	25.2	32.2	27.5
Fruit juices	33.2	8.7	25.1
Beans	19.3	23.0	20.5
Sancochos	16.3	26.6	19.7
Alimento natural	18.9	16.4	18.1
Spaghetti soups	12.6	16.2	13.8
Manioc and yam	7.9	21.8	12.5
Salads	15.0	4.8	11.6
Rice with other foods	7.5	9.3	8.1
Soup of rice	6.1	9.6	7.3
Spaghetti	8.3	5.0	7.2
Soups with different meats	7.5	5.4	6.8
Milk preparations for babies	7.7	4.3	6.6
Soup of corn	5.1	8.5	6.2
Soup of cereals	5.2	7.5	6.0
Mazamorras	3.6	7.2	4.8
Fish and seafood	3.0	7.1	4.4

Agua de panela = preparation with water and solid brown sugar; *arepas* = corn tortilla with or without cheese, sweet or salty; *sancocho* = stew, different types exist in the country; *alimento natural* = this category includes mother's milk; *mazamorras* = heavy soups prepared with corn or plaintains.

between the preparations used in the urban and in the rural areas: while coffee with milk, eggs, juices, salads and preparations made out of the home and more common in the urban areas, black coffee, *sancochos,* manioc and fish are more commonly consumed in the rural areas.

We selected the five first preparations used at breakfast, lunch and dinner and the followig conclusions may be drawn from table VIII: (a) Preparations made out of the home occupy the first and third places

Table VIII. Main preparations per region and type of meal (first five): Colombia 1981

Region	Breakfast	%	Lunch	%	Dinner	%
Atlantic	coffee with milk	62.2	soups	33.5	rice	51.8
	black coffee	46.2	rice	28.0	beef	42.8
	PMOH[1]	43.7	fruit juice	12.9	*plátano*	30.6
	plátano	33.6	PMOH	12.4	mixed rice	29.2
	manioc-yam	36.4	beef	11.3	PMOH	20.6
Eastern	broths	64.4	rice	41.9	rice	35.5
	PMOH	45.3	potatoes	33.7	*agua de panela*	34.4
	chocolate	37.8	*agua de panela*	23.1	potatoes	26.1
	black coffee	25.7	beef	18.3	beef	15.4
	coffee with milk	25.1	fruit juices	17.4	PMOH	15.2
Bogotà	PMOH	79.1	rice	58.8	rice	56.5
	chocolate	59.4	potatoes	40.8	potatoes	37.4
	broths	39.2	fruit juices	30.3	beef	21.5
	eggs	36.8	beef	26.6	PMOH	17.4
	coffee with milk	26.6	*agua de panela*	15.6		
Central	chocolate	79.7	rice	53.6	rice	65.0
	arepas[2]	67.5	*agua de panela*	34.7	beans	45.0
	PMOH	45.5	beef	23.5	*agua de panela*	37.8
	eggs	30.3	sancochos	22.6	beef	32.0
	agua de panela	28.8	*plátano*	17.0	*plátano*	22.6
Pacific	PMOH	48.6	rice	63.2	rice	59.6
	black coffee	37.7	sancochos	31.5	*plátano*	24.5
	chocolate	32.5	fruit juices	24.0	*agua de panela*	22.3
	eggs	24.0	potatoes	15.9	beef	21.3
	coffee with milk	19.0	beef	15.7	black coffee	18.1

[1] Preparations made out of the home (PMOH) include bread, cookies, biscuits and soda pop.
[2] *Arepas* = corn tortilla with or without cheese. Different kinds of *arepas* exist within the country.

among the preparations consumed at breakfast in all the regions. (b) Some sort of beef is consumed at breakfast in the Atlantic, Pacific and Central regions. (c) Rice is an important staple for lunch and dinner in all the regions. (d) Some sort of beef is used for lunch and dinner in all 5 regions. (e) Soups and sancochos are important at lunchtime.

VIII. Discussion

Herewith it is important to discuss and point out some of the conclusions of the analysis of the data presented above. It is understood that the DNP DANE PAN DRI survey performed in 1981 was not intended to be exhaustive but it represents one of the first efforts to study the feeding habits of the national population according to region, area and level of income.

This understanding allows the identification of problem areas in food habits.

(1) The foods most commonly used are, in order: onion, rice, potatoes, beef, vegetable oil, liquid milk, plaintains, panela, tomatoes and sugar. There are regional variations as follows: there is a higher consumption of potatoes in the Eastern region; of milk, bread and soda pop in Bogotá, of plaintains in the Pacific region, of rice, beef and manioc in the Atlantic region, and of *panela* and corn in the Central region.

(2) According to the data, rice is the food that contributes the most calories in the diet of the Colombian people. It is followed by plaintains, potatoes, corn and vegetable oil. The foods that contribute most protein are beef, rice, liquid mild and potatoes.

(3) It was found that there is a larger number of families in the urban areas in respect to the rural areas, with better adequacy in the average of proteins consumed, also with a better average in the consumption of calcium and vitamin A.

(4) Concerning the main preparations consumed, those are: rice (73%), preparations made out of the home (62.4%), chocolate (48.5%), *Agua de panela* (43.5%), and recipes with some sort of beef (43.1%). Other preparations with higher use are black coffee, plaintains, coffee with milk, eggs, potatoes, *arepas,* broths and fruit juices. The following are outstanding food practices: (a) preparations made out of the home are important, especially in the urban areas; (b) black coffee and coffee with milk is used in all the regions – in some regions there is a higher consumption of chocolate and *agua de panela;* (c) the consumption of preparations with some

sort of beef is higher in the Atlantic region, and lower in the Eastern region; (d) Potatoes are consumed more frequently in Bogotá and the Eastern region while the use of plaintains is more widespread in the Atlantic and Pacific regions; (e) soups are widely used in all the regions and areas.

(5) According to the type of meal (breakfast, lunch or dinner) the following comments are to be kept in mind: (a) preparations made out of the home are important for breakfast in all the regions; (b) a high consumption of rice exists at lunch and dinner in all the regions; (c) some kind of beef is always at use for lunch and dinner in all the regions; (d) black coffee and/or chocolate are widely used for breakfast while fruit juices are consumed more for lunch and *agua de panela* for dinner; (e) the main roots and starches used in the diet of Colombians are potatoes and plaintains.

The relationship between the government's agricultural policies and nutritional problems has been overlooked by planners and programers. Although the country has had a development in the agricultural and industrial sectors over the last years, the demand for better and cheaper foods has been increasing as the amount of foods that can be bought with a salary decreases. Programs on nutritional and consumer education have not covered the areas with the lowest socioeconomic status and programs like PAN (Programa del alimentación y nutritión) have been cancelled without examining their impact or performing an evaluation. Housing and living conditions of a large segment of the Colombian population make cooking conditions and food storage worse considering the lack of potable water and the cost of energy for cooking.

As some other scholars have pointed out, the causes of malnutrition are to be explained in the production, distribution and consumption systems and in the understanding and improvement of economic agricultural policies [16, 17]. More integrative efforts by the different governmental agencies are needed in order to solve the nutritional problems of the population. Planning of the system of production, distribution and consumption of foods should be made for short- and long-term solutions. The analysis of surveys such as the one presented above identifies problem areas where programers must start working.

IX. Summary

This paper describes one of the analyses of the last National Food Consumption Survey (DANE DNP PAN 1981). It focuses on the foods most commonly used and the calorie and protein contribution of these foods, and it considers the differences in rural and urban patterns of food consumption. It also explains the relationship between food consumption patterns and the government's agricultural policy. Nutritional problems require, for their understanding and solution, a comprehensive perspective of the economic, social and cultural issues.

References

1 Escobar, I.; Rodríguez, E.; Romero, M.E.; Escobar, I.: Hábitos alimentarios de la población Colombiana. Análisis de la Encuesta DANE, DNP DRI PAN U. Javeriana. FEI, Bogotá (1985).
2 Romero, M.E.: Hábitos alimentarios en Colombia. Bibliografía comentada. Bogotá (1983).
3 Romero, M.E.: Traditional foods of urban migrants to Bogotá; in Pelto, Vargas, Perspectives on dietary change. UNAM (MIT Press, Cambridge, in press).
4 Suescun, J.; Ortiz, N.: Los hábitos alimentarios como factores determinantes de la situación nutricional. Primer Congr. Soc. Colombiana de Tecnologia de Alimentos, Bogotá 1982.
5 Young, N.: Programa nacional de investigación en tecnologia de alimentos y nutrición. Colciencias, 1er Congr. Soc. Colombiana de Tecnologia de Alimentos. Bogotá 1982.
6 Tavera, J.; et al.: Nutrición y desarrollo económico. ICBF (Dirección de Nutrición. Bogotá 1970).
7 Pérez, M.P.; León, R.: Un enfoque microeconómico acerca del consumo de alimentos (Facultad de Economia, U. Andes, Bogotá 1981).
8 Machado, A.: Situación actual y tendencias de la industria de alimentos. 1er Congr. Soc. Co. Tecnología de Alimentos, Bogotá 1982.
9 Duque, G.: El sector agricola; en Nueva frontera, No. 493, pp. 9–10, 32–33 (Bogotá 1984).
10 Gonzalez, E.: Empleo; en Economia Colombiana No. 172–173 (Bogotá 1985).
11 Coyuntura Agropecuaria: CEGA, Bogotá 1984.
12 Douglas, M.: Fundamental issues in food problems. Current anthropology, vol. 25, pp. 498–499 (Chicago 1984).
13 Pardo, F.: Encuesta nacional de alimentacion nutrición y vivienda DANE DNP DRI PAN. La situación socioalimentaria de la población colombiana (DNP, Bogotá 1983).
14 ICBF, Instituto Colombiano de Bienestar Familiar: Tabla de composición de Alimentos Colombianos. s.f.
15 The National Research Council: Recommended dietary allowances; 9th ed. (Washington 1980).
16 Morales de Look, A.: Descripción y análisis de las políticas de alimentación y nutrición en Colombia (1974–1984); tesis FEI, Universidad Javeriana, Maestria en Alimentación y Nutrición, Bogotá (1985).
17 Chonchol, J.: Pour une stratégie alimentaire centrée sur les paysans (Centre d'études de T. Monde, Paris 1985).

Dr. Maria Eugenia Romero, Independent Researcher and Consultant,
Apartado Aereo 39372, Bogota D.E.1 (Colombia)

Wld Rev. Nutr. Diet., vol. 55, pp. 58–97 (Karger, Basel 1988)

Opioid Peptides from Food (the Exorphins)

Eugenio Paroli

Institute of Medical Pharmacology and Drug Addiction Center,
University of Rome 'La Sapienza', Rome, Italy

Contents

Coeliac Disease, Schizophrenia and Opioid Peptides

The discovery that opioid peptides are released by the digestion of certain food has followed a line of research that assumes pathogenic connections between schizophrenic psychosis and diet.

Remote assumptions of this theory exist in the observations of English and Italian physicians. The Italian psychiatrist Buscaino [1933] reported on the atrophy of the mucosa of the small intestine and the damage to the intestinal villus in up to 86% of his schizophrenic patients. And Paulley, the English pathologist, who first described the histological lesions of coeliac disease, wrote 'in contrast to ulcerative colitis (coeliac patients) showed extreme obsessional neuroses, suffering delusions, frequently believing they had cancer. Paranoid ideas were also frequent and many were considered psychotic or near psychotic' [quoted by Dohan, 1976].

Since the end of the 1960s, a certain number of observations have been accumulating on the frequent association between schizophrenic psychosis and coeliac disease, a gluten-sensitive enteropathy which affects predisposed subjects fed on cereals containing gliadin. Among others, the worsening of psychotic symptoms after substituting wheat gluten for soy flour, and on the other hand, the most rapid recovery after withdrawal of both gluten and milk proteins from the diet [Dohan et al., 1969; Dohan and Gransberger, 1973]; epidemiological data of the greatest occurrence of schizophrenia among women on a diet of cereals rich in gliadin [Dohan, 1966]. In the experience of several researchers, the recurring observation is that the presence of milk in the diet worsens the psychopathological symptoms of schizophrenia, as it would appear to substitute the negative effects of gluten in some patients put on a diet of cereals containing proteins different from gliadin [Singh and Kay, 1976]. The reasons for such similarities in certain constitutional and psychopathological symptoms between gluten enteropathy and schizophrenia are still undergoing research [Horrobin, 1979]. The responsibility of a protein or a peptide, at the same time entero- and psychotoxic, has prompted many studies, among which some developments in the opioid peptides field, which are the subject of the present survey. However, the immunopathological theory of coeliac disease also acknowledges some solidarity between this pathological condition and the schizophrenia, given the greater frequency (approximately 3 times) with which schizophrenics display antibodies to gluten proteins.

Several pioneering attempts to produce alterations in behaviour through the oral administration or intracerebral injection of gluten proteins to animals [Hemmings et al., 1977; Hemmings, 1978] had the merit of indicating the possibility of direct psychotoxic effects of gluten peptides.

As it was in those years that the aptitude of endogenous opioid peptides was shown to produce in rats, behavioural symptoms, like catatony, frequent in schizophrenics [Bloom et al., 1976], interest in the pathogenic role of certain diets concentrated on the possible alimentary contribution of absorption of opioid-like peptides from the gut [Klee et al., 1978; Zioudrou and Klee, 1979]. However, the role of endorphins in schizophrenia is still waiting to be clarified. Naloxone, which is the best-known antagonist, may alleviate hallucinations, but it is basically ineffective against the course of the disease [Watson et al., 1978]. On the other hand, while β-endorphin acts somewhat as a neuroleptic [Jaquet and Marks, 1976], the removal of an apparently abnormal β-endorphin-like peptide, should improve the psychotic state [Palmour et al., 1977]. Whatever the future developmet of research on this topic, it has had the merit of stimulating the search for potentially psychotomimetic peptides which are released by food. The presence of opioid peptides in pepsin digests of gluten or casein is widely recognized, and has been the object of review [Wajda et al., 1976; Klee et al., 1978; Zioudrou and Klee, 1978, 1979; Henschen et al., 1981; Morley, 1982; Moscucci, 1984; Brantl, 1986a, b]. Added to the extraordinary ascertainment which gives certain foods a further role in regulating the functions of the organism, numerous studies, relative to the source, have followed, on production mechanism, absorption, physiological and pharmacological properties, together with the attempts to derive more active and enzyme stable terms.

Opioid peptides from food have been called *exorphins* 'since they are exogenously derived substances with morphine-like activity' [Zioudrou and Klee, 1978]. On the other hand, the finding of an opioid peptide from β-casein digests, the first one whose sequence has been characterized, has introduced the term *β-casomorphin* [Brantl et al., 1979] to mean its source.

The Source of Exorphins

The principal sources of exorphins recognized so far are wheat gluten proteins [Klee et al., 1978; Zioudrou and Klee, 1978; Zioudrou et al., 1979], and their different classes, saline-soluble, gliadin and glutenin [Huebner et al., 1984]; fresh and dried cow's milk [Brantl and Teschemacher, 1979; Svedberg et al., 1985]; fresh cow's milk contaminated with caseolithic bacteria [Hamel et al., 1985]; casein [Meisel, 1986]; casein from human milk [Yoshikawa et al., 1983, quoted by Yoshikawa et al., 1984], casopeptone obtained from cow's milk [Wajda et al., 1976; Brantl and Teschemacher, 1979; Chang et al., 1985; Meisel, 1986]; bovine α-Sl casein [Zioudrou et al., 1979; Loukas et al., 1983]; buffalo milk β-casein [Petrilli et al., 1984]. According to some reports [Stan et al., 1983; Yoshikawa et al., 1984], also peptic digests of κ-casein, including the human one, contains opioid peptides. The elucidation of the primary structure of the major components of casein of different animal species, as well as human ones, has made it possible to pick out the sequence, or a similar characteristic sequence of an already isolated peptide, in the primary structure of casein, and to prepare by synthesis, and to test for opioid activity or opioid receptor affinity, such a fragment, whose release in the digestion of casein may be anticipated. A computer search on a known primary protein sequence is now being developed. Initially, it was acknowledged that the opiate peptide Tyr-Pro-Phe-Pro-Gly-Pro-Ile [Henschen et al., 1979a], corresponded to the sequence 60–66 of bovine β-casein (the first peptide to be called β-casomorphin or β-casomorphin 1–7). This sequence is present in the β-casein of other species like ovine and buffalo milk [Richardson and Mercier, 1979; Petrilli et al., 1984]. These caseins are, therefore, foreseeable sources of exorphins. Similarly, in human milk casein, the sequence Tyr-Pro-Phe- that makes up the N-terminus of a heptapeptide similar to bovine β-casomorphin 1–7, that is: Tyr-Pro-Phe-Val-Glu-Pro-Ile (human β-casein 51–57) was isolated. This peptide, directly obtained by synthesis, has shown opioid activity [Yoshikawa et al., 1984; Brantl, 1985; Koch et al., 1985]. With the same approach, followed by direct resort to chemical synthesis, as human casein primary structure was made available, the 'unnatural' amidated tetrapeptide fragment analogous to the potent 'natural' bovine β-casomorphine 1–4 amide (morphiceptin) has been proved to have opioid activity. Such a direction quickly tends to increase our notions on the existence of many exogenous opioid peptides and imposes great perception in admitting to new endorphin-like substances, so that, more than the presence in the body, no proof has been given of the system of biosynthesis and deposit. Recently [Yoshikawa et al., 1986], the opioid properties of one human β-casein fragment, with a different fundamental sequence from the one so far examined in β-casein (Tyr-X-Phe), as Tyr-X1-X2-Phe, has been isolated. Such a peptide reveals opioid properties and has been called *β-casorphin.* It corresponds to residue 41–44 of human β-casein, being apparently exclusive to human beings. Two other peptides, one of which with the sequence present in the α-lactoglobulin of various animal species, whose amide has been called *α-lactorphin* (Tyr-Gly-Leu-Phe), the other with the sequence present in bovine β-lactoglobulin (Tyr-Leu-Leu-Phe), both show opiate-like activity.

The approach to obtain regulatory peptides analysing the primary structure of important proteins is only just beginning. With regard to milk, Berg et al. [1986a, b] have isolated in the colostrum of *Camelus dromedarium,* subjected to peptic hydrolisis, fragments of β-casein showing sequences containing alternating proline residues resembling the structure of the opioid β-casomorphins.

Meat has also recently been singled out as a possible source of opioid peptides of alimentary origin. From bovine blood digested with gastrointestinal enzymes [Brantl et al., 1985; Brantl, 1986a, b], two types of opioid peptides were obtained, respectively, with Tyr-Pro-Phe-Trp-Ile sequence, similar to the one of bovine β-casomorphins, the other with Tyr-Pro-Trp-Thr-Gln sequence. A computer search for the respective parent proteins showed the former to correspond to the fragment 345–349 of mitochondrial cytochrome *b*, and it was called *cytochrophin;* the latter, particular in its structure, corresponds to the fragment 34–38 of the β-chain of bovine haemoglobin and was called *hemorphin.* As for protein cereals, the difficulty in establishing the sequence, and, therefore, of reproducing authentic peptides by synthesis, leaves some questions unresolved, with regard to either the type of sequence of opiopeptides, or the occurrence of both stimulatory, not naloxone-reversible, and opioid-like activity of peptides obtained from the digests of zein and hordein [Zioudrou and Klee, 1978, 1979].

Release of Exorphins

In vitro Release

Usually, the procedure consists in subjecting the food, milk for example, or relatively purified proteins (as in the case of casein or gluten proteins) to the effect of pepsin at pH of 2.5 for 30 min to 1 h. This procedure is sufficient to release exorphins and β-casomorphins. In some cases, digestion was made more complete with the added step-by-step of other enzymes to the substratum, like trypsin to the pepsin hydrolysates of glutenin and gliadin [Huebner et al., 1984], of pancreatic enzymes and intestinal peptidases to the hydrolysed peptics of milk [Svedberg et al., 1985], likewise for other enzymes like trypsin, chymotrypsin, elastase, carboxypeptidase A, leucine aminopeptidase [Petrilli et al., 1984], with the object of 'simulating' in vitro the pattern of digestion of the opioid peptides present in the pepsin digest in the gastro-intestinal tract, both to test the resistence of the peptide opiates present in the hydrolysate to other pancreatic and intestinal enzymes. Making the procedures of extraction, gel filtration to separate fragments between 100 and 1,800 daltons, chromatography and further separation and purification by reverse-phase and ion-exchange high-performance liquid chromatography (HPLC), the peptides can be isolated and tested for their opioid activity. Among the biological assays employed, the inhibition of electrically induced contractions of isolated myentheric plexus/longitudinal muscle of guinea-pig ileum (GPI; possibly rich in opiate μ-receptors) and mouse vas deferens (MVD; rich in δ-receptors). The inhibition of the rate of adenosine 3′:5′-cyclic phosphate (cAMP) accumulation in intact neuroblastoma-glioma hibrid cell lines [Sharma et al., 1975], which contains μ- and δ-opiate receptors [Yu et al., 1986], is an opiate test shown to be sensitive to gluten exorphins [Zioudrou et al., 1979]. The affinity for the subpopulation of opiate receptors is tried by binding inhibition assays against specific μ-, δ- or κ-receptor ligands on cell membranes of brain homogenates. More frequently, opioid peptides from milk have been obtained from extraction processes carried out on casopeptone or other commercially available preparations of digested casein. In the first cases of isolation of gluten or casein peptides, the peptide nature of opioid activity was confirmed by the disappearance after hydrolysis with pronase. Such a specification was justified by reports on the presence of a morphine-like substance of non-proteic nature in milk [Hazum et al., 1981]. There are no records devoted to assessing the yield in opioid peptides of pepsin digests. However, some data is available. Zioudrou et al. [1979] obtained as much as 28 μg of the more active opioid fraction from 100 g of gluten; among more recent reports [Chang et al., 1985], from 10 g of a casein digest, a total of 18 μmol of β-casomorphin was obtained, of which 43 nmol of the strongest morphioceptin, corresponding respectively to about 15 mg and 23 μg. Of course, the yield of opioid peptides from food depends both on the quantity of the substratum and the duration of hydrolysis. For example, testing the effect of caseolithic bacteria on cow's milk [Hamel et al., 1985], β-casomorphin immunoreactivity was shown, in some samples, after many day's incubation. Concerning the production in vitro of exorphins of cows' milk (more widely investigated), the digest contained essentially β-casomorphin 1–7, with its C-terminus hydrolysis products β-casomorphin 1–6, 1–4. It is interesting to note that β-casomorphin 1–7 is a bitter-tasting peptide six times more

Table I. Sources of exorphins an sequence of peptides identified in hydrolysed cereal products, casein or digested milk or casein products

Source	Name	Sequence	Author
Gluten	gluten exorphin	not determined	Klee et al. [1978]
Gliadin and glutenin hydrolysed by pepsin and/or trypsin	many peptide fractions	not determined	Huebner et al. [1984]
α-Casein	casein exorphin	not determined	Zioudrou, Klee [1978]
Cow α-casein pepsin hydrolysate in vitro	α-casein exorphin	Arg-Tyr-Leu-Gly-Tyr-Leu-Glu	Loukas et al. [1983]
		Arg-Tyr-Leu-Gly-Tyr-Leu	Loukas et al. [1983]
Cow β-casein peptone	β-casomorphin (1–7)	Tyr-Pro-Phe-Pro-Gly-Pro-Ile	Brantl et al. [1979]
	β-casomorphin (1–5)	Tyr-Pro-Phe-Pro-Gly	Henschen [1979a]
	β-casomorphin (1–3)	Tyr-Pro-Phe	Lottspeich et al. [1980]
Buffalo β-casein (59–68)	Pro-casomorphin	Val-Tyr-Pro-Phe-Pro-Gly-Pro-Ile-Pro-Lys	Petrilli et al. [1984]
	peptide B	Tyr-Pro-Phe-Pro-Gly-Pro-Ile-Pro	Petrilli et al. [1984]
Cow casein enzymatic hydrolysate in vitro	β-casomorphin (1–4) NH_2	Tyr-Pro-Phe-Pro-NH_2	Chang et al. [1985]
	8-prolil-β-casomorphin (1–7)	Tyr-Pro-Phe-Pro-Gly-Pro-Ile-Pro	Chang et al. [1985]
Human duodenal chyme at various times after the ingestion of fresh milk	mainly β-casomorphin (1–7)-like material *immunoreactivity*	not determined	Svedberg et al. [1985]
Duodenal chyme from minipigs fed with cow casein	β-casomorphin (1–11)	Tyr-Pro-Phe-Pro-Gly-Pro-Ile-Pro-Asn-Ser-Leu	Meisel [1986]
Cow's milk inoculated with caseolytic microorganisms	β-casomorphin (1–4) β-casomorphin (1–6) β-casomorphin (1–7) *'immunoreactivity'*	as reported above	Hamel et al. [1985]

bitter than caffeine, and is credited to give cheese its taste quality, followed by a heptapeptide starting from the amino acid Pro^2 of β-casomorphin 1–7 (Pro-Phe-Pro-Gly-Pro-Ile-Pro = fragment 61–67 of bovine casein) which is four times more bitter than caffeine [Shinoda et al., 1986]. In the enzymatic digestion of α-Sl bovine casein, fragments 90–96 and 90–95 are released (for the sequence, see table I).

As for the seasonal variations on the content of opioid peptides from cow's milk, the major yields of β-casomorphin were obtained in May and in October [Brantl and Teschemacher, 1979]. The enzymatic mixture adopted to obtain the release of opioid peptides form β-casein deserves some interest. In fact, not only does the output in the peptide sequences of casomorphins differ, depending on the enzymatic step of proteolysis [Svedberg et al., 1985], but the end-product obtained may sometimes show immunoreactivity, but not chemical identity with peptides extracted by pepsin digests, possibly being parent peptides of theirs [Petrilli et al., 1984; Svedberg et al., 1985; Chang et al., 1985].

Release into the Alimentary Canal

Svedberg et al. [1985] in human beings, and Meisel [1986] in male adult Göttingen minipigs have searched for the presence of β-casomorphins in the duodenal chyme at various intervals, after the administration of milk and acid-precipitated casein, respectively, on an empty stomach. In the tests on human beings, considerable quantities of β-casomorphin 1–7 immunoreactive material was present in the duodenal content: however, only a small part co-eluted with authentic β-casomorphin 1–7, the majority were eluted previously. This portion is thought to be made up of one sequence greater than β-casomorphin, having an additional amino acid at the N-terminus (the antibody used interrelated specifically at the C-terminus). Accordingly, opioid activity is lacking due to the absence of Tyr on the N-terminus. From similar conclusions come tests on the duodenal contents of the Göttingen minipig where, however, a peptide was found with the sequence of β-casomorphin 1–11 (Tyr-Pro-Phe-Pro-Gly-Pro-Ile-Pro-Asn-Ser-Leu), which being supplied with Tyr to the N-terminus, shows opiate activity in a receptor-binding assay. As stated before, even in the digestion in vitro of casein, Chang et al. [1985] reported a C-terminus extended peptide (8-prolyl-β-casomorphin 1–7) and Petrilli et al. [1984] a decapeptide with an extra amino acid at the N-terminus and two amino acids at the C-terminus which they termed procasomorphin. Apparently, the integrated action of proteolythic enzymes, as usually occurs in vivo, is able to release longer sequences of casein and, as an added bonus, it allows peptides longer than β-casomorphin 1–7 to survive hydrolysis. Some of these have the required structure for opioid activity, others may accomplish it in a more caudal intestine or, perhaps, in blood or elsewhere.

Relationship between Sequence and the Opioid Activity of Exorphins

Before the discovery of the exorphins of gluten and casein, and the sequencing of β-casomorphin [Henschen et al., 1979a; Lottspeich et al., 1980], the opioid peptides known may be brought back to the basic sequence of the Met- and Leu-enkephalin (Tyr-Gly-Gly-Phe-:Met5 or Leu5). Opioid peptides released from food proved that opioid activity may be carried out by very different sequences previously unknown. The α-Sl casein releases [Loukas et al., 1983] the sequences of the fragment 90–96, Arg-Tyr-Leu-Gly-Tyr-Leu-Glu, and of the fragment 90–95. The bovine β-casein released [Henschen et al., 1979a] fragments with the characteristic sequence Tyr-Pro-Phe-Pro-Gly-Pro-Ile (fragment 60–66) and, apparently, the amidated tetrapeptide Tyr-Pro-Phe-Pro-NH2, morphiceptin [Chang et al., 1985]. Human β-casein predictably releases similar peptides with the sequence Tyr-Pro-Phe-Val-Glu-Pro-Ile [Brantl, 1985], with the exception of the amidated analogue of morphiceptin, valmuceptin, because of the absence in human β-casomorphin 1–5 of the C-carboxy-terminal Gly, the amidation codon. Added to these 'natural' peptides are those coming from the digestion of the bovine blood, already mentioned, the cytocrophins 1–4, 1–5 [Brantl et al., 1985], with the sequence Tyr-Pro-Phe-Thr-Ile, and the haemorphins 1–4, 1–5 [Brantl et al., 1986], Tyr-Pro-Trp-Thr-Gln. Other sequences possibly released in the digestion of β-casein and α- and β-lactoglobulin of human and bovine milk have been recorded above. Tables I and II give the sequences of the chief opioid peptides obtained from the digestion of some food. Tables III and IV compare the opioid potency of 'natural' peptides obtained from the digestion of casein and gluten. Considering that the available data come from different experiences, the comparison between peptides and these and the endorphins and the opiates is often not available. Referring to the type of tests from a receptorial point of view, it appears that gluten exorphins are 100 times more active than morphine on the MVD [Zioudrou et al., 1979], and that compared to other fractions the gliadine fractions give peptides with

Table II. Sequences of human casomorphins according to primary sequence of human β-casein (realization by chemical synthesis)

Name	Sequence	Author
β-Casomorphin (1–7)	Tyr-Pro-Phe-Val-Glu-Pro-Ile	Brantl [1985]
β-Casomorphin (1–5)	Tyr-Pro-Phe-Val-Glu	Brantl [1985]
β-Casomorphin (1–4)	Tyr-Pro-Phe-Val	Brantl [1985]
β-Casomorphin (1–4) NH_2 (not 'natural')	Tyr-Pro-Phe-Val-NH_2	Yoshikawa et al. [1984]
(β-Casein 59–63)	Tyr-Gly-Phe-Leu-Pro	Yoshikawa et al. [1986]
β-Casorphin	Tyr-Pro-Ser-Phe-NH_2	Yoshikawa et al. [1986]
α-Lactorphin	Tyr-Gly-Leu-Phe-NH_2	Yoshikawa et al. [1986]

greater affinity for opiate receptors, when sensitivity is measured in mg peptide equivalent to 1 n*M* of morphine in bonding to opioid-active sites [Huebner et al., 1984]. The exorphins from α-Sl bovine casein, as a complex, and as authentic peptides of the fraction 90–96 of casein, present a relative affinity for δ-opiate receptors, as deduced from the lesser molar concentration required to inhibit the binding of [^{3}H]DALAMID[[^{3}H]Tyr]-D-Ala^2-Met-enkephalinamide [Loukas et al., 1983]. On MVD, a preparatus rich in δ-receptors, α-S1 casein exorphins would appear many times less active than morphine. The opiopeptides of both bovine and human β-casein have more affinity with μ-receptors, compared to δ-receptors. Although β-casomorphin 1–7 is approximately 300 times less potent than normorphine and Met-enkephalin, and 100 times less than morphine on guinea-pig ileum (GPI), it displays a 5-fold greater preference for μ-receptors compared to morphine, when the affinity for μ- and δ-receptors, respectively, is compared on a radioreceptor-binding assay [Brantl et al., 1981; Chang et al., 1981]. With the most potent morphiceptin, potency on GPI is the same as for morphine; however, the preference for μ-receptors is approximately 1,000 times greater [Chang et al., 1981]. The affinity of about 1,000 times for μ-receptors has made morphiceptin one of the elective ligands in the radioceptor assay procedures in the analyses of receptorial affinity of opiates or opiopeptides. An interesting example is the study of the receptorial affinity of benzomorphan opiates [Osborne and Herz,

Table III. Comparative opioid activity of food opiopeptides

Peptide	GPI	MVD	Author
Gluten exorphin	–	2*	Zioudrou and Klee [1978]
Casein exorphin	–	8,000*	Zioudrou and Klee [1978]
Cow β-casomorphin (1–7)	55.0	> 500	Henschen et al. [1979]
Cow β-casomorphin (1–5)	5.0	35	Henschen et al. [1979]
	6.5 (0.6)	42.1 (5.9)	Brantl et al. [1981]
Cow β-casomorphin (1–4)	21.90 (2.6)	84.3 (12.8)	Brantl et al. [1981]
Cow β-casomorphin (1–3)	> 200	–	Lottspeich et al. [1980]
Cow β-casomorphin (1–4) NH_2	0.13	15	Chang et al. [1985]
Human β-casein derived peptides			
β-Casomorphin (1–4)	27.60	–	Kock et al. [1985]
β-Casomorphin (1–5)	13.50	–	Kock et al. [1985]
β-Casomorphin (1–7)	29.0	–	Kock et al. [1985]
Cow casein-derived peptides			
90–96 α-casein	–	30	Loukas et al. [1983]
90–95 α-casein	–	70	Loukas et al. [1983]
Morphine	0.1	0.9	Chang et al. [1981]

GPI = Myentheric plexus – longitudinal muscle of guinea-pig ileum; MVD = mice vas deferens preparations. IC_{50} as μmol/l. (SD); *as ng/ml.

1983], where morphiceptin was used to cover μ-receptors, so as to reveal the affinity of the test compound by displacement.

On that occasion, the bi-phase course of the curve of competitive displacement of the selective bond with μ-receptors that characterizes morphiceptin was proved. Such behaviour matches the unique affinity of this peptide for μ-receptors to high (μ1) and low (μ2) affinity, with a preference of 30–50% for the first ones [Villiger, 1984]. Recourse to an opiate antagonist like naloxazone, which irreversibly and selectively blocks the μ-sites of high affinity, confirmed the special receptorial μ1 affinity of morphicep-

Table IV. Relation between structure and activity of β-casomorphins in binding assay ^{125}I-labelled enkephalin analogues (IC_{50}, μmol/l which reduced the binding of labelled compound by 50%). From Chang [1981]

β-Casomorphin	^{125}I-Labelled FK33824	^{125}I-Labelled (D-Ala2, D-Leu5) enkephalin
(1–7)	1.8±0.6	15.0
(1–5)	0.5±0.1	25
(1–4)	1.2±0.1	70
(1–4) NH_2	0.019±0.004	30
Morphine	0.0004±0.0002	0.035±0.005

tin, on which analgesic strength would depend (see further). Many of the morphine-like effects with casomorphins, would, in fact, depend on μ1 receptors, i.e. supraspinal analgesia, catalepsy, prolactin (PRL) release, acetylcholine turnover [Gavril et al., 1986]. Much of the information on the properties of morphiceptin is indirectly derived from its use as a μ-receptorial ligand. This said property determines the behaviour of the peptide towards other opiate tests [Day et al., 1981]. Interestingly, too high μ-agonist affinity opposes antagonism to ε-receptors (β-endorphin) [Day et al., 1981]. As for the potency of the more recently isolated peptides, human β-casomorphins are usually more alike for μ-receptors, with the exception of β-casomorphin 1–5, and are generally characterized by a low relative potency, compared both to bovine β-casomorphins, and endorphins or opiates. Cytochrophins and haemorphins are the least strong peptides, but they confirm the affinity for μ-receptors. The predictiveness of functional effects of receptorial affinity, as pointed out by Goldstein and James [1984], is limited. Indeed, the kinetics of opioids in the body is unrelated to the site of the receptors towards which they behave as preferential ligands. However, in short, it may be said [Chang and Cuatrecasas, 1981] that μ-receptors prevail in the hypothalamus and thalamus, whereas δ-receptors prevail in the brain cortex and in the striatum. Moreover, high concentrations of μ-receptors are found in areas that are relevant to pain sensation or analgesia production, such as the substantia nigra, the periaqueductal grey, the median raphe, the dorsomedial thalamus and the IVth layer of the cortex. Areas found to be rich in δ-receptors are parts of the limbic system associated with emotions and reward behaviour. δ-Receptors are present as well in the spinal cord, in substantia gelatinosa, where they may provide analgesic activity. Apparently, μ-receptors mediate many neuroendocrine functions, whereas δ-receptors prevail in peripheral tissue, particularly gastro-intestinal mucosa and intestine. It is rather interesting that the β-casomorphins are, for now, the unique 'natural' ligands of μ-receptors. As weill be said later on, the distance of a single amino acid from the aromatic ones Tyr- and Phe- is mandatory for prevailing μ-binding but other groups determine opioid potency as in morphiceptin [Chang et al., 1983]. Loew et al. [1986] have thoroughly investigated both the conformational and the electronic requirements in giving the terms of the morphiceptin series μ-receptor affinity. On the coordination complexes of amino acid sequences of β-casomorphin 1–5 and its fragments see Formicka-Kozlowska et al. [1984].

Absorption and Fate of Exorphins

When exorphins were shown to resist pepsin digestion and, later, that of other proteolithic enzymes, trypsin, chymotripsin, carboxypeptidase A and B, leucine-aminopeptidase, it was interpreted as predictive of their possibility of extensive absorption by the alimentary tract. Later, the resistance to pronase, a non-specified mixture of various endopeptidase (serine and metalloprotein) and exopeptidase from *Streptomyces griseus,* it was thought significant confirmation of the exorphins' resistance to hydrolysis [Brantl and Teschemacher, 1979]. It soon became evident that the enzymes specific for splitting peptides at the proline C-side bond were able to cleave the β-casomorphins [Hartrodt et al., 1982a, b; Kreil et al., 1983]. In the presence of the same enzyme in the plasma, it was attributed the short half-time, less than 10 min, of the β-casomorphins incubated with human bovine or rat plasma. In the intestinal tract, the brush border peptidases, which include dipeptidylaminopeptidase IV, also called dipeptidyl-peptidase IV, oligopeptidase, aminopeptidase A, glutamyl-transpeptidase and carboxypeptidase, split peptides into smaller peptides. They are present in rat foetus at the start of the differentiation, then reach a maximum, right after birth, and then drop to adult levels [Auricchio et al., 1978, 1981]. Other tissues, like placenta, submaxillary glands, kidney and meconium, of a full-term human new-born contain it [Caporale et al., 1985]. Dipeptidyl-peptidase IV splits the sequence of β-casomorphins on the C-side of the proline residue, releasing the dipeptides Tyr-Pro; Phe-Pro; Gly-Pro, with Ile [Hartrodt et al., 1982a]. In spite of this, when the decapeptide procasomorphin, found in proteolytic digest of milk, is incubated with the human meconium enzyme, it was cleaved only at the N-terminus Val-Tyr, the residue octapeptide practically surviving further hydrolysis. So, the procasomorphin has been proposed as a partly peptidase-resistant precursor, able to carry the active casomorphin moiety in the blood, and deliver it to the tissues [Caporale et al., 1985]. Consequently, a precursor with a somewhat longer sequence than the opioid ones released by in vitro digests of β-casein seems to be a candidate for the absorption of β-casomorphins. Accordingly, after taking milk or dairy products, in human beings, no immunoreactivity to the β-casomorphins 1–4, 1–5 or 1–7 was found in plasma [Teschemacher et al., 1986]. On the other hand, in the plasma of a calf, fed on colostral milk within 6 h of birth, a peptide was found with the immunoreactivity to the β-casomorphin 1–7 that survived hydrolysis when incubated with plasma [Umbach et al., 1985]. Reports on the presence of β-casomorphin plasma immunoreactivity have already been recorded. Worthy of mention is the observation of volunteers, especially when affected by lactose intolerance, who display IgG antibodies towards casein and β-casomorphin 1–5 [Cunningham-Rundles, 1986]. The absorption from the gut of peptides of greater molecular weight than of exorphins is not unexpected. Motivated by the nutritional theory of schizophrenia, research has been performed in the past that does not exclude this possibility [Hemmings et al., 1977; Hemmings, 1978], although the absorption of protein amino acids as oligopeptides [Silk, 1974] is considered the chief mechanism for absorption through the non-amino acid 'peptic route' [Ganapathy and Leibach, 1982]. This pattern supposes a transport site on the outer surface of the membrane of epithelial cell and a hydrolithic site inside the membrane. As for enkephalin or enkephalin-like sequences, no ileal [Kerchner and Geary, 1983] and non-oral [Kaghi and Lee, 1986] active transport systems of intact pentapeptides or their tetrapeptide moieties have been proven. Likewise, by the study of short-circuit current of rabbit ileum (Isc), no evidence of absorption of natural β-casomorphins has been observed [Hautefeuille et al., 1986]. However, some β-casomorphin derivatives, especially β(D-Ala^{2,4}; Tyr^{5}) casomorphin 1–5-NH2, which resisted hydrolysis, reduces the Isc when applied to mucosa, the effect being antagonized by serosal naloxone. This behaviour has been considered predictive of the absorption of the unbroken peptide. As for the large peptides with physiological relevance, some absorption by mucosae rich in peptidases is known, as for example with insulin [Fujii et al., 1985]. So, more insight on the absorption kinetics of peptides proposed as 'pro-casomorphin' is necessary. Indeed the proline-rich sequences in caseins are degraded with a somewhat relatively long half-life within eukaryotic cells [Rogers et al., 1986]. Another important aspect of the fate of exorphins in the organism concerns the penetration of the central nervous system (CNS) and the cerebrospinal fluid (CSF), taking into account the role of regulatory peptides that they, possibly, play. It has been reported [Dohan et al., 1978] that polypeptides obtained by peptic-pancreatic digestion of wheat gliadin, when injected by the intracerebral route in rats, induce stereotyped behaviour and catalepsy that resemble some schizophrenic symptoms. Ermisch et al. [1983] showed that in rats [^{3}H]β-casomorphin 1–5, injected in the carotid artery, mainly accumulates in the brain regions where the blood-brain barrier shows the more permeable capillaries: choroid plexus of the IVth ventricle, pineal gland, pituitary neurointermediate lobus (circumventricular organs) and the anterior pituitary. In the other regions the labelled β-casomorphin 1–5 is distributed evenly, although showing a minimum, but significant, uptake. As to the half-time of elimination [Stark et al., 1986], after intracerebraventricular injection, in brain stem and corpus striatum, radioactivity falls to a fifth in 10 min. The chief metabolites were tyrosine and phenylalanine, and various oligopeptides. The presence of these fragments is apparently the product of brain proline-directed proteases, mainly proline endopeptidase [Andrews et al., 1982; Orlowski, 1983]. This enzyme, which is capable of hydrolysing synthetic substratum on the carbonyl site of proline, is present in rats in a vast number of tissues, as well as in the brain and pituitary. The enzyme cleaves the peptidyl-prolyl-peptide and peptidyl prolyl-amino acid bonds. From the typical substratum X1-AA-Pro-X2 it releases X2. Apparently, the rat prolyl-endopeptidase is identical to the 'postproline-cleaving enzyme' first discussed by Walter et al. [1971]. Its physiological role appears to be relevant since many hormonal or active peptides, such as oxytocin, vasopressin, angiotensin II, bradykinin and others, are split by it. Being a serine protease, prolyl-endopeptidase is inactivated by DFP. However, the distribution and half-time of regulatory peptides in the brain is not a problem peculiar to exorphins. Peptide transport systems have not been documented so far for the blood-brain barrier. Thus, it appears unlikely that distribution of peptides in the brain occurs after systematic absorption, whereas peptides absorbed into the CSF may rapidly distribute into the systematic circulation and exit from the brain at the arachnoid villi [Partridge, 1984]. Neither may a leaky or immature blood-brain barrier be further maintained, with the exception of an early foetal stage [Cornford et al., 1982; Cornford and Cornford, 1986], although the penetration rate of insulin in rat brain reaches the low adult level 10–15 days postpartum. According to Partridge et al. [1981] and Partridge [1983, 1984], regulatory peptides are able to enter the brain through circumventricular organs, choroid plexus, the median eminence and the organum vasculosum of the lamina terminalis of the hypothalamus, the subfornical organ at the roof of the third ventricle. The ependyma of these regions is likewise permeable. From here, the peptides easily reach their receptors on the neurons of the median eminence, or go to other adjoining regions of the brain, and penetrate the CSF. As far as the other areas are concerned, the brain microvessels contain numerous peptide-splitting enzymes [Palmieri et al., 1985]. To overcome this, it was suggested [Partridge, 1983, 1984] that receptors for regulatory peptides were present on the luminal side of the blood-brain barrier, and that a 'second-messenger' system in the endothelium processes the peptide signal. The binding of various transmitters has so far been shown by the brain's microvessels; among others, of insulin [Pillion et al., 1982], α-adrenergic and β-adrenergic agonists and histamine [Nathanson and Glaser, 1979; Peroutka et al., 1980].

Physiological and/or Pharmacological Effects of Exorphins

Much research has been carried out in order to establish the effects of food ingestion which releases exorphins or of the exorphins themselves on body functions, deducing the *opiate* character from antagonism by naloxone. Some of this research was centred on the functions and behaviour of suckling animals, since milk is a well-recognized source of exorphins. Other, numerous studies were intended to discover the 'properties' of exorphins, by means of experimental devices to better compare the intensity of effects to the dose, regardless of any nutritional purpose but with resort to naloxone, to check the opioid mechanism of the effects. Moreover, with the purpose of bypassing hydrolysis, resort to enzyme-resistant derivatives has often been done. Finally, a great deal of study, only briefly mentioned here, refers to exorphins, because of their unique receptor affinity in radioligand binding assays. Though employing exorphins as a tool for research, these studies have given a better insight on the mechanisms of food opiopeptide in modifying organic functions.

Gastro-Intestinal Motility

Pepsin-hydrolysed gluten prolongs the time of intestinal transit in human beings, according to controlled research by Morley et al. [1983]. The effect is prevented by the contemporary oral administration of 2 mg of naloxone. Other research has reached somewhat different conclusions [Corazza et al., 1984], where undigested gluten prolonged, by a naloxone-sensitive mechanism, mouth to caecum transit time, while a fraction of peptic-tryptic digest of gluten increased the intestinal transit time, but the effect was naloxone-resistant. Several studies have been carried out on the effect of milk opiopeptides on intestinal motility. The intraduodenal instillation of casopeptone (2 ml/kg), or β-casomorphin 1–5 (0.5 mg/kg) did not modify the regular trains of slow waves that characterize the intestinal movements in cats [Wienbeck et al., 1985]. Transit time was slowed down in mice, after a stable analogue of the β-casomorphin 1–4, Tyr-Pro.NMe-Phe-DPro-NH2(PLO17), administered intracerebroventricularly, subcutaneously and intrathecally [Shook and Burks, 1986; Mailman, 1986]. It is interesting to note that the effect for the subcutaneous route could be antagonized by naloxonazine, a μ1 receptor irreversible antagonist, but not by naloxone. Furthermore, the inhibition of intestinal transit, accomplished by PLO17, seems to occur at different levels, central and peripheric, only the former being resistent to naloxonazine. On the isolated ileum of guinea-pig, β-casomorphin 1–5 induces a temporary and reversible suspension of the ongoing peristaltic activity, increasing the duration of the contraction-free intervals. The effect is stereospecifically antagonized by naloxone, and bears out the presence of μ-opioid receptors, which with κ-receptors, appear to be mainly responsible for the opioid control of peristaltic periodicity [Kromer et al., 1980]. An increase of contractions of the small intestine was observed in dogs with morphiceptin, as with other opioids, after intra-arterial administration [Burks et al., 1982]. However, as for intestinal motility, differences in response to opiates among the animal species have been pointed out [Ambinder and Schuster, 1979]. The absorption process of electrolytes which characterizes an important component of the antidiarrhoeal effect of opiates, is enhanced in vitro only by some synthetic analogues of β-casomorphin 1–5, stable to enzymatic hydrolysis [Hautefeuille et al., 1986]. At variance, natural enkephalins are active on the same experimental device.

Gastric Acid Secretion

Gastric acid secretion, either basal or pentagastrin-stimulated, is inhibited in dogs with gastric fistulas by the administration of β-casomorphin 1–5, an effect antagonized by naloxone [Smagin et al., 1983]. A pepsin digest of κ-casein by intracerebroventricular administration stimulates the stomach's acid secretion, apparently with a naloxone-sensitive mechanism. The accountable peptide is unknown [Stan et al., 1983]. No inhibiting activity of hydrolysed gluten [Morley et al., 1983], casopeptone [Klier et al., 1982] or β-casomorphins [Schick et al., 1984] was observed on gastrin secretion. Differently with morphine, FK33824 and (D-Ala^2, Met^5) enkephalinamide, the β-casomorphin 1–5, by tube or intraperitoneally, does not protect rats against gastric haemorrhagic lesions produced by absolute ethanol or HCl [Paroli et al., 1986]. However, morphiceptin exerts some cytoprotection when given at higher doses (5 μM/kg) per os, and D-Ala^2-morphiceptin also intraperitoneally (ED_{50} 2 μM/kg) [Paroli et al., unpublished data]. Other observations have reported the reduction, non-naloxone reversible, of the intestinal absorption of rhamnose after hydrolysed gluten ingestion in volunteers [Morley et al., 1983]. Epithelial differentiation of the small intestine explants from suckling rats, 1–3 weeks from birth, appears to be stimulated by morphiceptin [Pasquarelli and Paroli, 1986]. Jejunum explants showed an increase of maltase activity, concentration-related, when incubated with morphiceptin (10^{-6} to 10^{-3} M). As compared to controls, the increase is significant up to the second week of life. Later, morphiceptin failed to increase maltase activity, as at the weaning stage, adult levels are reached.

Appetite

Although opiopeptides play a role in the feeding process, the opioid receptors involved, apparently of κ-subtype [Gosnell and Lipton, 1986] are not stimulated by gluten exorphins [Morley et al., 1983]. Indeed, when rats are injected intracerebroventricularly with morphiceptin, food intake does not increase, and it actually decreases. According to Morley et al. [1984], the μ-opioid receptors could mediate an anorexic mechanism. A possible role of exorphins in *allergic reactions* to milk and, possibly, to other foods, needs to be thoroughly investigated. Morphiceptin, as well as other opioids and opiates, produces skin wheal and flare, due to degranulation of mastocytes [Casale et al., 1984]. It may be remembered that a fragment of human β-casein, Val-Glu-Pro-Ile, is common to human β-casomorphin and to an immune-stimulating hexapeptide obtained from milk digests

[Parker et al., 1984]. However, membrane receptors for casein on human neutrophils, lymphocytes and monocytes, to which chemotactic effect of casein or milk is reported, do not bind β-casomorphin 1–5 [Lewis and Van Epps, 1983].

Hormone Release

Outstanding physiological studies concern the role of exorphins on the secretion of endocrine pancreas. On the pancreas of dogs, isolated and perfused [Ipp et al., 1978], and the isolated islets of Langerhans of rats [Green et al., 1980], the opiate alkaloids, as well as the endogenous opiopeptides, are able to modify the function, probably through a specific opioid receptorial mechanism, in accordance with immunocytochemical studies. Possibly, the effect of the opioid substances consists in the initial suppression of the inhibiting influence of somatostatin, followed by an increase in the secretion of insulin and glucagon. Naloxone alone, however, is able to increase the sole insulin secretion [Ipp et al., 1978]. On the other hand, when the availability of insulin is compromised, as in type I diabetes, opiates and opiopeptides determine hyperglycaemia, due to the unbalanced stimulation of glucagon [Ipp et al., 1980]. Schusdziarra et al. [1981], after the intragastric instillation of peptic digest of gluten in dogs, observed the augmentation of the postprandial rise of plasma insulin and glucagon. Intragastric, but not intravenous naloxone, reduced the response of the insulin, but intensified that of glucagon to digested gluten. Apparently naloxone works through an as yet unknown local mechanism, not requiring systematic absorption to function. On the other hand, the dissociation between the two hormonal responses was explained as being due to the primary depression of the insulin secretion, a hormone that inhibits the secretion of glucagon. However, in the isolated rat islets, β-casomorphin 1–5, at variance with a typical μ-ligand as DAMME, raised the release of insulin in a naloxone-reversible way [Nieter and Schatz, 1981; Nieter et al., 1981]. Accordingly, in dogs, the oral administration of casopeptone, or the addition to a test meal enriched with saccharose, of a mixture of β-casomorphins 1–7, 1–5, 1–4 and β-casomorphin 1-4-NH2, intensified the rise of plasma insulin by 6 or 7 times, an effect blocked by naloxone. The oral administration of met-enkephalin [Schudziarra et al., 1983a] behaved in the same way. Worthy of mention is the observation that in dogs, the i.v. infusion of β-casomorphin does not modify the basal (fasting) level of insulin, but raises it if the insulin release is prestimulated by the infusion of a solution of amino acids and glucose. The outstanding effects are shown by the β-casomorphin 1–4, limitedly opioid, while its amidated derivative, morphiceptin, was inactive, as were other synthetic derivatives with high opioid potency. The effects of natural β-casomorphins on plasma insulin were prevented by naloxone [Schusdziarra et al., 1983d]. There is reason to believe that the raise in postprandial plasma insulin produced by exorphins is restrained by the contemporary rise in the release of somatostatin, a peptide thought to be implicated in the so-called 'nutrient homeostasis'. In dogs, both fresh milk and the mixture of natural β-casomorphins (12 mg) added to a meal containing saccharose induce a naloxone-reversible rise of plasma somatostatin immunoreactivity. The i.v. infusion of β-casomorphin 1–5 also raises the somatostatin in plasma, when its release is prestimulated by a mixture of glucose and amino acids [Schusdziarra et al., 1983b]. Hydrolysed gluten and, to some extent, undigested gluten also raise somatostatin plasma immunoreactivity in man [Morley et al., 1983]. The effect is antagonized by naloxone. The exorphins apparently act in a different way from opiates and endorphins in regulating glycaemia. On fasting, there is no difference in behaviour: the plasma level of somatostatin is lowered [Schusdziarra et al. 1983c] and that of insulin is not stimulated by the casomorphins. As the glycaemia rises after a meal, the exorphins augment plasma insulin and, at the same time, plasma levels of glucagon and somatostatin. On the contrary, Leu- and Met-enkephalin seem to inhibit the release of insulin and somatostatin, even after the prestimulation with amino acids and glucose, apparently by activation of δ-receptors [Schusdziarra et al., 1983b, d]. One is tempted to speculate that the endogenous opioid system contributes to lessen glucose utilization and to initiate feeding, whereas exorphins maximize (by acting on peripheral, mainly δ, opiate receptors) the utilization of carbohydrates from food, and possibly restrain feeding behaviour by releasing somatostatin. Similar response shows the *pancreatic polypeptide* whose basal level is decreased by exorphins, while being raised by hydrolysed gluten, casopeptone or by a mixture of β-casomorphins as it raises the blood level of nutrients, once more, in the opposite way from endogenous opioids [Schusdziarra et al., 1983e]. Both effects of exorphins are antagonized by naloxone.

As occurs with other opiates, morphiceptin, injected intracerebroventricularly (ICV) induces the release of PRL and somatotrophin (GH), the former by a receptorial μ1 mechanism (blocked by naloxazone), while the release of GH appears as a δ-receptor effect [Koenig et al., 1983]. Furthermore, morphiceptin inhibits the release of luteinizing hormone (LH) in ovariectomized rats [Leadem and Kalra, 1985]. β-Casomorphin 1–7, in the same way, raises plasma level of PRL in male adult rats by i.p. (15 mg/kg) according to Nedvidkova et al. [1985]. Pertinent to this is research set up to establish if the absorption of opioids during lactation could modify the composition of secreted milk [Hossain et al., 1986]. In lactating rabbits, milk did not undergo any variation in composition, with the exception of some modification in the electrophoretic mobility of proteins. Besides, for the continuous taking of opioids, β-casomorphins included, the protein content and RNA with mRNA were reduced in the mammary gland. The effect was prevented by naloxone. A dose of 0.3 mg/kg, i.v. of β-casomorphin inhibits the release of thyrotropin-releasing hormone (TRH) by the hypothalamus, decreases the thyroid-stimulating hormone (TSH) plasma level while preventing cold-induced TSH elevation, due, apparently, to a dopaminergic mechanism which is antagonized by haloperidole [Mitsuma et al., 1984]. Although performed according to a pharmacological model, these studies may foresee a role for food opiopeptides in cold acclimatization. However, lipolysis is stimulated by endogenous opiopeptides but not by exorphins, in rabbit isolated adipocytes [Richter and Schwandt, 1985].

Behavioural Effects

Motor Behaviour. Catalepsy was reported to follow the intracerebral administration of a polypeptidic subfraction of gliadin, by Dohan et al. [1978], although the opioid mechanism remains unsettled. When β-casomorphins became available, it was observed that an ICV injection produced an 'unusual cataleptic state', at variance with opiates and endorphins, not associated with rigidity. Ligand-binding assays showed that β-casomorphin 1–4 was able to compete with the binding of [^{3}H]spiroperidol [Brantl et al., 1980]. Naloxone prevented the cataleptic state, but only partially. In other studies β-casomorphin 1–4 was found to be five times more potent than morphine in producing a state of rigidity, recorded as a tonic activity in the electromyogram, when injected directly into the striatum of unanaesthetized rats [Havemann and Kuschinsky, 1981]. The effect is antagonized by naloxone and appears to be related to high-affinity μ1-receptors situated 'post-synaptically' on dopaminergic neurons. The behavioural response typified by immobility contains both components, opioid and antidopaminergic (neuroleptic-like), according to Traber et al. [1984]. Intracisternally injected, morphiceptin depresses motor behaviour in concurrence with indirect evidence of a fall of dopamine release in the striatum [Yonehara and Clouet, 1984]. As the opiopeptides of β-casein can modulate the dopamine turnover in the brain, it also appears from research by Galimberti et al.

[1986], that the oral administration of morphiceptin to 4-day-old rats determined a decrease in dopamine metabolites (DOPAC) in the hypothalamus. The structure of β-casomorphin, on the other hand, seems to contain dopaminergic as well as opioid potentialities. Working on metabolically stable derivatives of β-casomorphin 1–5, Rauca and Matthies [1986] observed that in rats the unilateral striatal micro-injection of β-casomorphin-like agents produces contralateral rotations, as does apomorphine, while the relative *des*-Tyr derivatives behave like neuroleptics as far as the direction of movement is concerned. Since the N-terminus tyrosine determines the affinity for opiate receptors, so much so that the compounds *des*-Tyr are not analgesic, a shift of the affinity towards dopamine receptors may be expected if splitting of β-casomorphins by aminopeptidase occurred. Connected to the activity on motor behaviour is the inhibition by morphiceptin (infused in the lateral ventricle) of lordosis behaviour induced in female rats primed with oestrogen and progesterone, during mating [Pfaus et al., 1986]. An effect facilitating the lordosis by high doses of morphiceptin is explained by dependence on μ1 high-affinity receptors of inhibition, while μ2 receptors, with low bonding affinity with morphiceptin, would mediate facilitation of this behaviour. Also lower organisms change motor behaviour in the presence of β-casomorphins. The β-casomorphin 1–5 exerts an inhibitory effect on movements of insects (*Scotia segetum* Schiff and *Periplanata americana* L.) according to Otto et al. [1982]. Under the effect of the intra-abdominal injection of β-casomorphin insects showed changes in their habitual behaviour, a play of antenna and lateral contortions of the limbs, diminished touch reactivity and, finally, hypotonia and tiredness. The isolated ganglion of *Scotia segetum* Schiff reacts to β-casomorphin 1–5, with a sharp increase in spontaneous bio-electric activity. However, in rats, the morphiceptin (μ-agonist) does not determine by ICV, mioclonic contractions, as observed with δ-agonists [Haffmans and Dzoljic, 1983]. As a whole, exorphins influence motor behaviour like opiates do, but inhibitory effects are prevailing and these are apparently supported by a modulation of dopamine receptors. Possibly, the availability of β-casomorphins has thrown light on that receptor subpopulation which is implicated in catatony and sedation by opiates. The experimental models do not foresee effects of β-casomorphins on motor behaviour occurring after the ingestion of milk or dairy products. However, neurochemical effects were observed in the first few days after birth, so the hypothesis that exorphins may play a role in the behaviour of the *suckling* new-born [Zioudrou and Klee, 1979] cannot be dismissed.

Analgesia, Tolerance, Dependence. The demonstration of the analgesic effect of β-casein opiopeptides was quickly provided after isolation. However, only ICV analgesia can be shown, so that it is doubtful whether this can be reproduced by foods that release these casein exorphins; if not, perhaps in the first days after birth. It is a well-known medical prescription to provide the narcotic-addicted new-born with 'frequent feedings to reduce restlessness' [Berkow and Talbott, 1977]. β-Casomorphin, 1–5 is about 10–20 times less potent than morphine [Brantl et al., 1981; Matthies et al., 1984]. The β-casomorphin 1–4-NH2, morphiceptin, is, in its turn, about 10 times stronger than the latter [Chang et al., 1982]. As mentioned, this increase in the analgesic strength of morphiceptin is attributed to the particular disposition of the same to bind itself to the high-affinity μ1-receptors (see above). Accordingly, while naloxone is hardly effective in opposing morphiceptin analgesia, the μ1-irreversible blocking naloxazone completely annuls it [Zhang et al., 1981]. The selective affinity of β-casomorphins, particularly morphiceptin, to μ-receptors is apparently related to one single amino acid distance between the two aromatic amino acids, tyrosine and phenylalanine, at variance with the enkephalin structure, which is known to have more of an affinity with δ-receptors [DiMaio and Schiller, 1980]. In morphiceptin, C-terminus amidation intensifies the biological effects with peptides in which a neutral amino acid is followed by glycine. It is probable that the gastro-intestinal tract and pancreas have such an amidating enzyme [Chang et al., 1985]. However, the extent of the analgesic effects of β-casomorphins has prompted the realization of more active and enzyme-stable derivatives. As for 'natural' β-casomorphins, β-casomorphin 1–5 produces an appreciable analgesia by ICV, in the range of 0.06–2 μmol/rat. The weakest β-casomorphin, 1–7, produces, however, a more lasting analgesia, probably because of the transformation in the brain in a more active term. The N-terminus tripeptide of β-casomorphin is devoid of analgesic effect. As for the substitutions found effective in intensifying analgesia, with few exceptions they reflect resistance to splitting by proline endopeptidase (formerly postproline cleaving enzyme) and peptidyl-dipeptidase IV, as well as the lesser implicated aminopeptidase. However, affinity to binding sites is also modified. Amidation increases binding to μ-sites, whereas the introduction of D-Ala^2 increases δ-binding [Chang et al., 1981; Brantl et al., 1982]. The introduction of D-Pro and of D-Phe in proper positions increases μ-binding of β-casomorphin 1–5 and the substitution with *D*-pipecolic acid increases μ-binding in the brain and δ-receptor affinity in organ preparations [Liebmann et al., 1986]. Among the more efficacious substitutions is that of D-Pro for Pro^2 and Pro^4, with the substitution of N-MePhe or D-Phe^3 for Phe^3 in the sequence of β-casomorphins. Besides, *D*-pipecolic acid has been substituted for Pro^4 with an increase of analgesic activity. Considering that β-casomorphin 1–5 is 20 times less active than morphine given by ICV, not being at all analgesic when given i.v., the substitution with D-Pip^4 (depilorphin) and with D-Pro^4 (deprolorphin), in place of Pro^4, supplies two peptides, respectively 9.3 and 28 times more analgesic than morphine when taken ICV, whereas once and twice as potent as morphine by i.v. injection. The substitution with D-Pro^4 in morphiceptin (deproceptin), supplies a peptide 8 times more analgesic than morphine by ICV and 50% more potent than morphine when taken by i.v. injection [Matthies et al., 1982, 1984]. The further substitution of deproceptin at Phe^3 with N-$MePhe^3$ supplies a peptide (PLO17) about 35 times more analgesic than morphine when injected in the periaqueductal grey area and, therefore, by local infusion capable of inducing physical dependence and withdrawal symptoms after intraperitoneal naloxone injection [Chang et al., 1983]. Deprolorphin also induces tolerance to the analgesic effect as deduced by the cross-tolerance with morphine [Grecksch and Ruthrich, 1985]. The substitution with D-Ala^2 also increases the analgesic potency of morphiceptin. The duration of analgesia is also increased by the forementioned substitutions, in some cases by 5 or 6 times (by ICV). Since morphiceptin is a selected ligand for μ-receptors, and according to recent studies it should be the unique antagonist of the supposed endogenous antiopiate [Zadina and Kastin, 1985] peptide (Tyr-Pro-Leu-Gly-NH2 or Tyr-MIF-1), the question may be posed as to whether prolonged contact with morphiceptin by foods might modify the sensitivity to opiates of μ-receptors. Research on embryonic cells from rats' forebrain or hindbrain cultured with morphiceptin have shown, over time, a decrease in μ-receptors [Simantov et al., 1984]. As for other in vitro observations, the physiological implications are a matter of speculation. For example, tolerance to the synaptic influence of opiates is associated with down-regulation of δ-receptors, whereas that of μ-receptors after morphiceptin is minimal [Dingledine et al., 1983].

Psychobiological Implications. In foetal and new-born development, the population of μ-opioid receptors appears to precede that of other receptors, such as δ-receptors [Wohltmann et al., 1982]. Considering the alimentation rich in casein in the period after birth, the question has been raised as to whether the possible absorption of milk's opiopeptides may have some consequence on the development of the brain. The daily subcutaneous administration of morphiceptin in the 7 days following birth facilitates the maze test performance later carried out on the 25th day, and to a greater extent in females than in males [De Avila and Martinez, 1985]. A similar effect follows the administration of metenkephalin and DADLE(D-Ala^2, D-Leu^5 enkephalin). Mauk et al. [1982] have proved in rabbits that the injection in the rostral region of the fourth ventricle of morphiceptin disrupts a simple aversely motivated learned response, such as the classically conditioned eyelid/nictitating membrane response, without disrupting the unconditioned response. The effect is abolished by naltrexone. It is absent with D-Pro^2 substitute and is stronger than with (N-Me′Phe^3, D-Pro^4) morphiceptin. It has been suggested that μ-receptors, selectively sensitive to morphiceptin, are situated in the circuitry that codes the learned response, possibly localized in the periaqueductal grey/periventricular region of the IVth ventricle. According to another proposal, morphiceptin, and opiopeptides in general, work by eliminating conditioned fear. Lavond et al. [1983] observed in rabbits that the micro-infusion by the derivative (N-$MePhe^3$,D-Pro^4) morphiceptin disrupted the conditioned response (bradycardia) associated with a musical tone delivered simultaneously to periorbital shock. The removal of fear might explain the minimizing of *separation-induced distress* in chicks by β-casomorphins. Chicks put in isolation 3–5 days after hatching emit a distress vocalization whose rate can be measured. The intraventricular injection of β-casomorphins minimizes the distress reaction, β-casomorphin 1–5 being the most effective term. Apart from the systematic absorption of β-casomorphins, it has been suggested that opioid receptors in the gut can pick up the signal from β-casomorphins released by milk and transmit it as a signal of comfort and/or relief. Hence isolation distress vanishes [Panksepp et al., 1984]. It might be mentioned here that opiates are credited as acting as an internal cue in status-dependent learning [Colpaert, 1977]. Moreover, while the stimulation of the μ-receptor is an essential element in the genesis of the morphine-like discriminative effect of opioid peptides, morphiceptin, the main μ-agonist among β-casomorphins is minimally effective on this behaviour [Locke and Holtzmann, 1986]. No similar data is available on cereal exorphins, which makes up the chick's alimentary source, of this type of emotional control. Intriguing is the observation by Lindström et al. [1984] of the presence of opiopeptides containing proline with β-casomorphin characters in the CSF of nursing mothers with postpartum psychosis, a condition characteristically connected with the beginning of lactation.

This peptide fraction was also present in the plasma in some patients. Small quantities were present in the CSF, but not in the plasma of normal nursing mothers. The peptides were partially destroyed by the post-proline cleaving enzyme. The genetic polymorphism of β-casein admits the occurrence of β-casomorphin variants as the source of the above-mentioned peptides. It is, however, unknown whether casein is the source of these peptides, and if casein peptides with the sequence of β-casomorphin are re-absorbed in the course of their synthesis from the mammary gland.

Neurophysiological and Neurochemical Studies

The neurophysiological and neuropharmacological effects of opiopeptides of the exorphin series were examined at a synaptic level, especially at the level of neuronal circuits rich in opioid receptors. Among these is the formation of hippocampus and some romboencephalic formations. The synaptic transmission to hippocampus through the perforant pathway which monosynaptically connects the enthorinal cortex with the granule cells of the dentate area is depressed by the intracerebral injection of β-casomorphin 1–5 and, even more intensely, by the derivative deprolorphin [Reymann et al., 1985]. This peptide shows both a neuroleptic type of action [Matthies et al., 1984] as well as a reduction in (^{3}H)spiperone binding on dopamine and 5-HT receptors. In vitro it favours the release of dopamine from striatal slices, and the potassium-stimulated release of acetylcholine from hippocampus slices [Kammerer et al., 1985]. Furthermore, it causes a large enhancement of the population spike of CA1 pyramidal cells in vitro slice preparations [Reymann et al., 1983], showing an opposite behaviour when compared with electrical activity induced in dentate granule cells. Similar effects were observed in the CA1 hippocampal region with morphiceptin and its derivative PLO17 [Bostock et al., 1984]. Regarding the release of neurotransmitters, morphiceptin was only slightly effective in inhibiting the release of acetylcholine and norepinephrine evoked by electrical field stimulation in slices of rabbit hippocampus, a typical effect of opioids, now attributed to κ-agonists [Jackisch et al., 1986a, b]. Morphiceptin selectively inhibits the calcium-dependent action potential duration of a subpopulation (with δ-μ or μ-receptors, possibly coupled to potassium channels) of mouse dorsal root ganglion neurons grown in dissociated cell culture [Werz and Macdonald, 1983, 1985]. Thus, spinal analgesia could also be provided by morphiceptin. Morphiceptin influences *cardiovascular regulation,* as shown in several experimental preparations. Many of these studies have used casomorphins as tools for research, to ascertain the mediation of the effects by a proper subpopulation of opioid receptors. Morphiceptin, injected into the pressor area of the ventrolateral medulla in rats, reduces arterial bood pressure and slows down heart rate, but has the opposite effect if injected in the depressor area. Apparently the former injection decreases the sympathetic outflow, the latter inhibits the physiological restraint of sympathetic outflow [Punnen and Sapru, 1986]. The intrathecal injection of (N-MePhe^3, D-Pro^4) morphiceptin produces a decrease in the mean arterial pressure and heart rate which appears after a few minutes and reaches its maximum (a decrease of 50% of both) after 1 h. Intrathecal naloxone prevents the fall. On the other hand, the cardiovascular response to such a stable casomorphin-like compound is not prevented by vagotomy. Of the endogenous opiopeptides only dynorphin A intrathecally causes similar effects [Li and Han, 1984]. Cardiovascular response to opiopeptides is apparently a complex one. It depends on the contribution of many regulatory areas of the brain as well as on their peripheral effects on the heart and circulatory reflexes. As far as the areas of the CNS are concerned, only large doses of morphiceptin injected into the caudal region of the medial portion of the nucleus of tractus solitarius have been found to modify cardiac frequency [Hassen et al., 1982a, b].

In this region, opioid δ-receptors appear to be more involved. When injected in the periventricular region of the hypothalamus, morphiceptin and other opiopeptides, and the morphine itself, induced hypotension and bradypnea in high doses, like other μ-selected agonists [Faden and Feuerstein, 1983], but tachycardia in small doses (0.64–6.4 nmol). Worthy of mention is that the pre-existent state of the circulation dictates the response. As a matter of fact, ICV injected morphiceptin produces a rise in the heart rate and blood pressure level in spontaneously hypertensive, but not in normotensive, rats [Widy-Tyszkiewicz et al., 1986]. However, the intravenous administration of β-casomorphins is generally followed by a transitory slowing of cardiac frequency and short-lived hypotension [Wei et al., 1980], both graded according to their μ-receptor affinities. By this route, the same cardiocirculatory response is also observed in hypertensive rats. Cardiac slowing and hypotension are further reported in dogs after the intra-arterial injection of β-casomorphins [Sander and Giles, 1982]. As for the myocardic response to β-casomorphins, on the guinea-pig papillary muscle, β-casomorphin 1–5 increases the maximum upstroke velocity of slow action potential and produces oscillatory after-depolarizations resulting in spontaneous activity. The effect is explainable with an increase in calcium channel activity [Nilius et al., 1985]. On the matter of the effects of opioids on cardiovascular function, see Holaday [1983]. No less complex is the action of the morphiceptin and its stable derivative PLO17, on the respiratory functions. Differently from bradypnea from morphine, PLO17 induces is dogs an increase up to 60% of mean instantaneous minute ventilation associated with rapid and shallow breathing. Insufficient alveolus ventilation is inferred by the drop in pO_2 and the rise of pCO_2, possibly due to an increase in dead-space ventilation associated with the rapid and shallow breathing pattern observed. Only δ-agonists seem to induce periodic breathing [Schaeffer and Hadded, 1985]. In cats, the neural discharge from carotid body chemoreceptors is not influenced by μ-agonist opioids nor morphiceptin; a δ-agonist has been shown to be inhibitory [Kirby and McQueen, 1986]. However, J pulmonary receptors opiate stimulation [Sapru et al., 1981] appears to be μ-receptor-mediated, since morphiceptin injected into the external jugular vein of rats, as far as the right atrium, causes apneoa, bradycardia ad hypotension within fractions of seconds. The apnoea was followed by rapid and shallow breathing [Krieger and Sapru, 1985]. Interestingly, lungs are rich in proline endopeptidase [Behal et al., 1986]. These studies confirm growing evidence that the opioid receptors that mediate analgesia are rarely, if ever, involved in the respiratory depression pattern typically induced by opiates [Ling et al., 1983]. On the other hand, β-casomorphins so far investigated may induce, if absorbed in proper amounts, cardiorespiratory insufficiency. From a practical point of view, morphiceptin, as a μ-'natural' agonist may, perhaps, give a model to make new kinds of analgesic devoid of respiratory depression. Furthermore, the potential toxicity of morphiceptin on respiratory and cardiovascular functions could be blamed, where intestinal absorption allowed it, among the sudden death causes in children, as put forward by β-endorphin [Kuich and Zimmermann, 1981].

Conclusions

Since the first exciting reports on opioid activity of gluten and casein peptic digests, the opiopeptides derived from foods have been the subject of a great deal of study, as well as having been used as a tool for research in the field of opiate physiology and pharmacology. Noteworthy is the fact that fundamental foods, like milk and cereals, release during digestion peptides whose better-known physiological actions are explained on endocrine pancreas functions and intestinal motility. Furthermore, many experimental studies have explored food's opiopeptides in doses and by routes devoid of nutritional relevance. While they met the requirements of a better knowledge of biological properties of exorphins or related compounds, they put forward new hypotheses on the role that may be played by food opiopeptides on behaviour. Maze learning, conditioned fear, motor behaviour, pain perception, metabolic control, respiratory and circulatory responses all appear to be influenced by many exorphins, mainly by some β-casomorphins. Whether these influences may be expected on ingestion of the relative food is a matter of speculation. Many uncertainties surround the absorption by the gut and the uptake by the brain of food's opiopeptides. In spite of this, more than one possible progenitor peptide, particularly stable to enzymes, has been found in the gut or plasma after the ingestion of the supplying food. Furthermore, antibodies to β-casomorphin have been found in the plasma of some patients. According to some research, it may be expected that cerebral microvessels contain enzymes suitable for freeing the active sequence from the parent peptide, as well as the circumventricular regions of brain being allowed some exorphins to directly reach the neurons. Moreover, for many regulatory peptides the role of distribution on the signal delivery in the brain is to be clarified. Up to now, psychopharmacological experiences point to a role of exorphins on behaviour and possibly on stress protection, particularly in the first few days of life. It is, also, interesting that exorphins apparently favour the maturation of the alimentary canal, like the development of certain enzymes in the gut explants foresees, and that in the new-born animal they modulate the dopamine turnover in the brain. If the endorphin theory of schizophrenia expected an endorphin-like exogenous counterpart from cereals, the exorphins have filled this requirement. As for exorphin implication in schizophrenia, no single study aimed at this topic appeared. The presence of a β-casomorphin-like material in the CSF and plasma of postpartum psychosis deserves, however, further attention. Finally, the discovery of exorphins has provided the chemistry of opioid peptides of not less than, perhaps, eight new peptide sequences (bovine α-casein peptides, β-casomorphins and morphiceptin, human β-casomorphins, possibly, α- and β-lactoglobulin peptides, haemorphins and cytochrophins), until then unexpected. It is noteworthy that the unique μ1-receptor 'natural' ligand, morphiceptin, may be supplied by food and, supposing it is centrally available, that it exerts some discriminative stimulus effect of a morphine-like type. One can hardly accept the idea of a purposeless occurrence. Perhaps opioid receptor characterization in the organism may well throw some light on a somewhat food opioid system and on its role in nutrition.

Finally, unsuitable storage, as well as processing of certain foods, like milk, for instance, may allow patients with underlying illnesses of the gastro-intestinal tract and the pancreas to ingest an excessive charge of exorphins compared to the ability to handle them as far as their absorption and metabolism are concerned. In addition, the composition of the diet has been shown to enhance the binding of opioid peptides, as observed with a diet enriched with sucrose, in mice [Marks-Kaufman et al., 1986]. So a chance for engendering the kind of 'food addiction' outlined by Philpott et al. [1980] may ensue; in the meantime a non-immunogenic pattern of food allergy is likely to occur in these patients because of the mastocyte degranulating property of β-casomorphins. Apparently, due to the exorphins, also the subject of food poisoning and food intolerance may find new pathways to explore.

References

Ambinder, R.F.; Schuster, M.M.: Endorphins: new gut peptides with a familiar face. Gastroenterology *77:* 1132–1140 (1979).

Andrews, P.C.; Minth C.D.; Dixon J.E.: Immunochemical characterization of a proline endopeptidase from rat brain. J. biol. Chem. *257:* 5861–5865 (1982).

Auricchio, S.; Greco, L.; De Vizia, B.; Buonocore, V.: Dipeptidylaminopeptidase and carboxypeptidase activities of the brush border of rabbit small intestine. Gastroenterology *75:* 1073–1079 (1978).

Auricchio, S.; Stellato, A.; De Vizia, B.: Development of brush border peptidase in human and rat small intestine during fetal and neonatal life. Pediat. Res. *15:* 991–995 (1981).

Behal, F.J.; Baker, C.R.F., Jr.; Zolfaghari, R.: Human lung postproline endopeptidase. Fed. Proc. *45:* 678 (1986).

Beg, O.U.; Von Bahr-Lindstrom, H.; Zaidi, Z.H.; Jornvall, H.: Characterization of a camel milk protein rich in proline identifies a new β-casein fragment. Regulat. Peptides *15:* 55–62 (1986).

Berkow, R.; Talbott, J.H.: The Merk manual of diagnosis and therapy (Merck Sharp & Dohme Research Laboratories, Rathaway 1977).

Bloom, F.; Segal, D.; Ling, N.; Guillemin, R.: Endorphins: profound behavioral effects in rats suggest new ethiologic factors in mental illness. Science *194:* 630–632 (1976).

Bostok, E.; Dingledine, R.; Xu, G.; Chang, K.-J.: Mu opioid receptors participate in the excitatory effect of opiates in the hippocampal slices. J. Pharmac. exp. Ther. *231:* 512–517 (1984).

Brantl, V.; Neubert, K.: Novel opioid peptides derived from human β-casein: human β-casomorphins. Eur. J. Pharmacol. *106:* 213–214 (1985).

Brantl, V.: Opioid peptides derived from food proteins. Trends pharmacol. Sci. *16:* 6 (1986a).

Brantl, V.: Pharmacological effects of opioid peptides derived from food proteins. Int. Meet. Food Safety and Health Protection, 1986 (Consiglio Nazionale delle Ricerche, Roma 1986b).

Brantl, V.; Gramsch, C.; Lottspeich, F.; Henschen, A.; Jaeger, K.-H.; Herz, A.: Novel opioid peptides derived from mitochondrial cytochrome *b:* cytochrophins. Eur. J. Pharmacol. *111:* 293–294 (1985).

Brantl, V.; Gramsch, C.; Lottspeich, F.; Mertz, R.; Jaeger, K.H.; Herz, A.: Novel opioid peptides derived from hemoglobin: hemorphins. Eur. J. Pharmacol. *125:* 309–310 (1986).

Brantl, V.; Henschen, A.; Lottspeich, F.: Pharmacological properties of β-casomorphins. Arch. Pharmacol. *311:* R67 (1980).

Brantl, V.; Pfeiffer, A.; Herz, A.; Henschen, A.; Lottspeich, F.: Antinociceptive potencies of β-casomorphin analogs as compared to their affinities towards μ and δ opiate receptors sites in brain and periphery. Peptides *3:* 793–797 (1982).

Brantl, V.; Teschemacher, H.: A material with opioid activity in bovine milk and milk products. Arch. Pharmacol. *306:* 301–304 (1979).

Brantl, V.; Teschemacher, H.; Blasig, J.; Henschen, A.; Lottspeich, F.: Opioid activities of β-casomorphins. Life Sci. *28:* 1903–1909 (1981).

Brantl, V.; Teschemacher, H.; Henschen, A.; Lottspeich, F.: Novel opioid peptides derived from casein (β-casomorphins). Hoppe-Seyler's Z. physiol. Chem. *360:* 1211–1216 (1979).

Burks, T.F.; Hirning, L.D.; Galligan, J.J.; Davis, T.: Motility effects of opioid peptides in dog intestines. Life Sci. *31:* 2237–2240 (1982).

Buscaino, V.M.: Componenti enterogene della demenza precoce. Riv. pat. nerv. ment. *41:* 483 (1933).

Caporale, C.; Fontanella, A.; Petrilli, P.; Pucci, P.; Molinaro, M.F.; Picone, D.; Auricchio, S.: Isolation and characterization of dipeptidyl peptidase IV from human meconium. FEBS Lett. *184:* 273–277 (1985).

Casale, T.B.; Bowman, S.; Kaliner, M.: Induction of human outaneus mast cell degranulation by opiates and endogenous opioid peptides: evidence for opiate and nonopiate receptor participation. J. Allergy clin. Immunol. *73:* 775–781 (1984).

Chang, K.J.; Cuatrecasas, P.: Heterogeneity and properties of opiate receptors. Fed. Proc. *40:* 2729–2734 (1981).

Chang, K.J.; Cuatrecasas, P.; Wei, E.T.; Chang, J.K.: Analgesic activity of intracerebroventricular administration of morphiceptin and β-casomorphins: correlation with the morphine (μ) receptor binding affinity. Life Sci. *30:* 1547–1551 (1982).

Chang, K.J.; Killian, A.; Hazum, E.; Cuatrecasas, P.; Chang, J.K.: Morphiceptin (NH_2-Tyr-Pro-Phe-Pro-$CONH_2$): a potent and specific agonist for morphine (μ) receptors. Science *212:* 75–77 (1981).

Chang, K.J.; Su, Y.F.; Brent, D.A.; Chang, J.K.: Isolation of a specific μ-opiate receptor peptide, morphiceptin, from an enzymatic digest of milk proteins. J. biol. Chem. *260:* 9706–9712 (1985).

Chang, K.J.; Wei, E.T.; Killian, A.; Chang, J.K.: Potent morphiceptin analogs: structure activity relationships and morphin-like activities. J. Pharmac. exp. Ther. *227:* 403–408 (1983).

Colpaert, F.C.: Minireview: narcotic cue and narcotic state. Life Sci. *20:* 1097–1108 (1977).

Corazza, G.R.; Frazzoni, M.; Strocchi, A.; Prati, C.; Sarchielli, P. et al.: Alimentary exorphin action on motility and hormonal secretion of gastrointestinal tract. Dev. Neurosci. *18:* 243–247 (1984).

Cornford, E.M.; Braun, L.D.; Oldendort, W.H.; Hill, M.A.: Comparison of lipid-mediated blood-brain barrier penetrability in neonates and adults. Am. J. Physiol. *243:* C161–C168 (1982).

Cornford, E.M.; Cornford, M.E.: Nutrient transport and the blood-brain barrier in developing animals. Fed. Proc. *45:* 2065–2072 (1986).

Cunningham-Rundles, C.: Gastrointestinal absorption of β-casomorphin in humans. Clin. Res. *34:* 493A (1986).

Day A.R.; Freer, R.J.; Liao, C.S.: Morphiceptin (β-casomorphin (1–4)) amide a peptide opioid antagonist in the field stimulated rat vas deferens. Res. Commun. chem. Pathol. Pharmacol. *34:* 543–546 (1981).

De Avila, A.; Martinez, J.L., Jr.: Sex-dependent effects of neonatally administered morphiceptin and D-Ala-D-Leu-enkephalin on maze learning in rats. Behav. neural Biol. *44:* 393–402 (1985).

DiMaio, J.; Schiller, P.W.: A cyclic enkephalin analog with high in vitro opiate activity. Proc. natn. Acad. Sci. USA *77:* 7162–7166 (1980).

Dingledine, R.; Valentino, R.J.; Bostock, E.; King, M.; Chang, K.J.: Down-regulation of delta but not mu opioid receptors in the hippocampal slice associated with loss of physiological response. Life Sci. *33:* suppl. 1, pp. 333–336 (1983).

Dohan, F.C.: Cereals and schizophrenia; data and hypotesis. Acta psychiat. scand. *42:* 125–152 (1966).

Dohan, F.C.: The possible pathogenic effect of cereal grains in schizophrenia; celiac disease as a model. Acta neurol. *XXI:* 195–205 (1976).

Dohan, F.C.: Schizophrenia and neuroactive peptides from food. Lancet *i:* 1031 (1979).

Dohan, F.C.; Grasberger, J.C.: Relapsed schizophrenics: earlier discharge from the hospital after cereal-free, milk-free diet. Am. J. Psychiat. *130:* 685–688 (1973).

Dohan, F.C.; Grasberger, J.C.; Lowell, F.M.; Johnston, H.T., Jr.; Arbegast, A.W.: Relapsed schizophrenics: more rapid improvement on a milk and cereal-free diet. Br. J. Psychiat. *115:* 595–596 (1969).

Dohan, F.C.; Levitt, D.R.; Kushnir, L.D.: Abnormal behaviour after intracerebral injection of polypeptides from wheat gliadin. Pav. J. biol. Sci. *13:* 73–82 (1978).

Ermisch, A.; Ruhle, H.J.; Neubert, K.; Hartrodt, B.; Landgrat, B.: On the blood-brain barrier to peptides: ([^{3}H] β-casomorphin 5) uptake by 18 brain regions in vivo. J. Neurochem. *41:* 1229–1233 (1983).

Faden, A.I.; Feuerstein, G.: Hypothalamic regulation of the cardio-vascular and respiratory systems role of specific opiate receptors. Br. J. Pharmacol. *79:* 997–1002 (1983).

Feuerstein, G.; Faden, A.I.: Differential cardiovascular effects of m, d and k opiate agonists at discrete hypothalamic sites in the anesthetised rat. Life Sci. *31:* 2197–2200 (1982).

Formicka-Kozlowska, G.; Pettit, L.D.; Steel, J.; Hartrodt, B.; Neubert, K.; Rekowski, P.; Kupryszewski, G.: The coordination of copper (II) with β-casomorphin and its fragments. J. inorg. Biochem. *22:* 155–163 (1984).

Fujii, S.; Yokoyama, T.; Ikegaya, K.; Sato, F.; Yokoo, N.: Promoting effect of the new chymotrypsin inhibitor FK-448 on the intestinal absorption of insulin in rats and dogs. J. Pharm. Pharmacol. *37:* 545–549 (1985).

Galimberti, R.; Restani, P.; Volterra, A.; Cuomo, V.; Galli, C.L.; Racagni, G.: Effetto di una somministrazione acuta di β-casomorfine sul sistema dopaminergico del ratto neonato. XXIIIo Congr. Nazionale della Società Italiana di Farmacologia, San Remo 1986, abstr., p. 73 (ECIG, Genova 1986).

Ganapathy, V.; Leibach, F.M.: Peptide transport in intestinal and renal brush border membrane vesicles. Life Sci. *30:* 2137–2146 (1982).

Goldstein, A.; James, I.F.: Multiple opioid receptors criteria for identification and classification. Trends Pharmacol Sci. *5:* 503–505 (1984).

Gosnell, B.A.; Lipton, J.M.: Opioid peptide effects on food intake in the rabbit. Fed. Proc. *45:* 605 (1986).

Grecksch, G.; Ruthrich, H.L.: The analgesically effective peptide deprolorphin exhibits cross-tolerance to morphine in rats. Biomed. Biochim. Acta *44:* 649–652 (1985).

Green, I.C.; Perrin, D.; Pedley, K.C.; Leslie, R.D.G.; Pyke, D.A.: Effect of enkephalins and morphine on insulin secretion from rat islets. Diabetolgia *19:* 158–161 (1980).

Haffmans, J.; Dzoljic, M.R.: Differential epileptogenic potentials of selective mu and delta opiate receptor agonists. J. neural Transm. *57:* 1–11 (1983).

Hamel, U.; Kielwein, G.; Teschemacher, H.: β-Casomorphin immunoreactive materials in cow's milk incubated with various bacterial species. J. Dairy Res. *52:* 139–148 (1985).

Hartrodt, B.; Neubert, K.; Fischer, G.; Demuth, U.; Joshimoto, T.; Barth, A.: Degradation of β-casomorphin-5 by proline-specific-endopeptidase (PSE) and post-proline-cleaving enzyme (PPCE). Pharmazie *37:* 72–73 (1982a).

Hartrodt, B.; Neubert, K.; Fischer, G.; Schulz, H.; Barth, A.: Synthese und enzymatischer Abbau von β-Casomorphin-5. Pharmazie *37:* 165–169 (1982).

Hassen, A.H.; Feuerstein, G.Z.; Faden, A.I.: Cardiovascular responses to opioid agonists injected into the nucleus of tractus solitarius of anesthetized cats. Life Sci. *31:* 2193–2196 (1982a).

Hassen, A.H.; Feuerstein, G.; Pfeiffer, A.; Faden, A.I.: Delta versus mu receptors: cardiovascular and respiratory effects of opiate agonists into nucleus tractus solitarius of cats. Regulat. Peptides *4:* 299–309 (1982b).

Hautefeuille, M.; Brantl, V.; Dumontier, A.M.; Desjeux, J.F.: In vitro effects of β-casomorphins on ion transport in rabbit ileum. Am. J. Physiol. *250:* 692–697 (1986).

Havemann, U.; Kuschinsky, K.: Further characterization of opioid receptors in the striatum mediating muscular rigidity in rats. Arch. Pharmacol. *317:* 321–325 (1981).

Hazum, E.; Sabatka, J.J.; Chang, K.-J.; Brent, D.A.; Findlay, J.W.A.; Cuatrecasas, P.: Morphine in cow and human milk: could dietary morphine constitute as ligand for specific (μ) receptors? Science *213:* 1010–1012 (1981).

Hemmings, W.A.: The entry into the brain of large molecules derived from dietary protein. Proc. R. Soc. Lond. B *200:* 175–192 (1978).

Hemmings, C.; Hemmings, W.A.; Patey, A.L.; Wood, C.: The ingestion of dietary protein as large molecular mass degradation products in adult rats. Proc. R. Soc. Lond. B *198:* 439–353 (1977).

Henschen, A.; Lottspeich, F.; Brantl, V.; Teschemacher, H.: Novel opioid peptides derived from casein (β-casomorphins). II. Structure of active components from bovine casein peptone. Hoppe-Seyler's Z. physiol. Chem. *360:* 1217–1224 (1979a).

Henschen, A.; Lottspeich, F.; Brantl, V.; Teschemacher, H.: β-Casomorphins: exorphins derived from casein peptone; in Voelter, Struct. Act. Nat. Peptides Proc. Fall Meet. Ges. Biol. Chem. (1979b), pp. 211–224 (1981).

Holaday, J.W.: Cardiovascular effects of endogenous opiate systems. Annu. Rev. Pharmacol. Toxicol. *23:* 541–594 (1983).

Hook, V.Y.H.; Affolter, H.U.: Precursor form of carboxypeptidase H: a processing enzyme involved in the synthesis of peptide hormones. Fed. Proc. *45:* 1718 (1986).

Horrobin, D.F.: Schizophrenia: reconciliation of the dopamine, prostaglandin and opioid contents and the role of the pineal. Lancet *i:* 529–530 (1979).

Hossain, M.A.; Kumar, A.; Ganguli, N.G.: Impact of opioid administration on the compositional changes in milk and body tissues of lactating rabbits. Milchwissenschaft *41:* 273–277 (1986).

Huebner, F.R.; Lieberman, K.W.; Rubino, R.P.; Wall, J.S.: Demonstration of high opioid-like activity in isolated peptides from wheat gluten hydrolysates. Peptides *5:* 1139–1147 (1984).

Ipp, E.; Dobbs, R.; Unger, R.H.: Morphine and β-endorphin influence on the secretion of the endocrine pancreas. Nature, Lond. *276:* 190–191 (1978).

Ipp, E.; Schusdziarra, V.; Harris, V.; Unger, R.H.: Morphine-induced hyperglycemia: role of insulin and glucagon. Endocrinology *107:* 461–463 (1980).

Jackisch, R.; Geppert, M.; Brenner, A.S.; Illes, P.: Presynaptic opioid receptors modulating acetylcholine release in the hippocampus of the rabbit. Arch. Pharmacol. *332:* 156–162 (1986a).

Jackisch, R.; Geppert, M.; Illes, P.: Characterization of opioid receptors modulating noradrenaline release in the hippocampus of the rabbit. J. Neurochem. *46:* 1802–1810 (1986b).

Jacquet, Y.F.; Marks, N.: The C-fragment of β-lipotropin: and endogenous neuroleptic or antipsychotogen? Science *194:* 632–635 (1976).

Kaghi, S.D.; Lee, W.H.L.: Enkephalin hydrolysis in homogenates of various absorptive mucosae of the albino rabbit; similarities in rats and involvement of aminopeptidase. Life Sci. *38:* 2019–2028 (1986).

Kammerer, E.; Koch, S.; Leon Roque, D.M.: Action of (D-Pro^4)-β-casomorphin1–5 on processes of synaptic transmis. Biomed. Biochim. Acta *44:* 1379–1387 (1985).

Kerchner, G.A.; Geary, L.E.: Studies on the transport of enkephalin-like oligopeptides in rat intestinal mucosa. J. Pharmac. exp. Ther. *226:* 33–38 (1983).

Kirby, G.C.; McQueen, D.S.: Characterization of opioid receptors in the cat carotid body involved in chemosensory depression in vivo. Br. J. Pharmacol. *88:* 889–898 (1986).

Klee, W.A.; Zioudrou, C.; Streaty, R.A.: Exorphins: peptides with opioid activity isolated from wheat gluten, and their possible role in the etiology of schizophrenia; in Usdin, Bunney, Kine, Endorphins in mental health research, pp. 209–218 (Macmillan, New York 1978).

Klier, M.; Holland, A.; Schick, R.: Fuente, A. de la; Brantl, V.: Exorphins modulate postprandial insulin but not gastrin release in dogs. Acta endocr., Copenh. *246:* suppl., pp. 111–112 (1982).

Kock, G.; Wiedemann, K.; Teschemacher, H.: Opioid activities of human β-casomorphin. Arch. Pharmacol. *331:* 351–354 (1985).

Koenig, J.I.; Mayfield, M.A.; Krulich, L.: Differential control of prolactin and growth hormone secretion by the opiates. Fed. Proc. *42:* 301 (1983).

Kreil, G.; Umbach, M.; Brantl, V.; Teschemacher, H.: Studies on the enzymatic degradation of β-casomorphins. Life Sci. *33:* suppl. 1, pp. 137–140 (1983).

Krieger, L.; Sapru, H.N.: Morphiceptin (a mu opiate receptor agonist) stimulates pulmonary J receptors in the rat. Fed. Proc. *44:* 425 (1985).

Kromer, W.; Pretzlaff, W.; Woinoff, R.: Opioids modulate periodicity rather than efficacy of peristaltic waves in the guinea pig ileum in vitro. Life Sci. *26:* 1857–1865 (1980).

Kuich, T.E.; Zimmerman, D.: Could endorphins be implicated in sudden-infant-death syndrome? New Engl. J. Med. *304:* 973 (1981).

Lavond, D.G.; Mauk, M.D.; Madden, J., IV; Barchas, J.D.; Thompson, R.F.: Abolition of conditioned heart-rate responses in rabbits following central administration of N-MePhe^3, D-Pro^4 morficeptin^1. Pharmacol. Biochem. Behav. *19:* 379–382 (1983).

Leadem, C.A.; Kalra, S.P.: Effects of endogenous opioid peptides and opiates on luteinizing hormone and prolactin secretion in ovariectomized rats. Neuroendocrinology *41:* 342–352 (1985).

Lewis, S.L.; Van Epps, D.E.: Demonstration of specific receptors for fluoresceinated casein on human neutrophils and monocytes using flow cytometry. Inflammation *7:* 363–375 (1983).

Li, S.J.; Han, J.S.: Depressor and bradicardic effect following intrathecal injection of NMePhe^3, D-Pro^4 morphiceptin in rats. Eur. J. Pharmacol. *99:* 91–95 (1984).

Liebmann, C.; Szucs, M.; Neubert, K.; Hartrodt, B.; Arold, H.; Barth, A.: Opiate receptor binding affinities of some D-amino acid substituted β-casomorphin analogs. Peptides *7:* 195–199 (1986).

Lindström, L.H.; Nyberg, F.; Terenius, L.; Bauer, K.; Besev, G.; Gunne, L.M.; Lyrenas, S.; Willdeck-Lund, G.; Lindberg, B.: CSF and plasma β-casomorphin-like opioid peptidexin-postpartum psychosis. Am. J. Psychiat. *141:* 1059–1066 (1984).

Ling, G.S.F.; Spiegel, K.; Nishimura, S.I.; Pasternak, G.W.: Dissociation of morphine analgesic and respiratory depressant actions. Eur. J. Pharmacol. *86:* 487–488 (1983).

Locke, K.W.; Holtzman, S.G.: Behavioral effects of opioid peptides selective for mu or delta receptors. I. Morphine-like discriminative stimulus effects. J. Pharmac. exp. Ther. *238:* 990–996 (1986).

Loew, G.; Keys, C.; Luke, B.; Polgar, W.; Toll, L.: Structure-activity studies of morphiceptin analogs: receptor binding and molecular determinants of μ-affinity and selectivity. Mol. Pharmacol. *29:* 546–553 (1986).

Lottspeich, F.; Henschen, A.; Brantl, V.; Teschemacher, H.: Novel opioid peptides derived from casein (beta-casomorphins). III. Synthetic peptides corresponding to components from bovine casein peptone. Hoppe-Seyler's Z. physiol. Chem. *361:* 1835–1839 (1980).

Loukas, S.; Varoucha, D.; Zioudrou, C.; Streaty, R.A.; Klee, W.A.: Opioid activities and structures of α-casein-derived exorphins. Biochemistry, N.Y. *22:* 4567–4573 (1983).

Mailman, D.: Influence of naloxone and caerulein on D-Ala[2] metenkephalinamide effects on canine ileal absorption and motility. Fed. Proc. *45:* 1044 (1986).

Marks-Kaufman, R.; Hamm, M.; Barbato, G.F.: The effects of dietary sucrose on opiate receptor binding in lean and obese mice. Fed. Proc. *45:* 424 (1986).

Matthies, H.; Ruetrich, H.L.; Stark, H.; Barth, A.; Neubert, K.; Hartrod, B.: Deproceptin, Tyr-Pro-Phe-D-Pro-NHZ, another casomorphin with remarkable analgesic action. Acta biol. med. germ. *41:* K39–K42 (1982).

Matthies, H.; Stark, H.; Hartrodt, B.; Ruethrich, H.L.; Spieler, H.T.; Barth, A.; Neubert, K.: Derivatives of β-casomorphins with high analgesic potency. Peptides *5:* 463–470 (1984).

Mauk, M.D.; Madden, J., IV; Barchas, J.D.; Thompson, R.F.: Opiates and classical conditioning: selective abolition of conditioned responses by activation of opiate receptors within the central nervous system. Proc. natn. Acad. Sci. USA *79:* 7598–7602 (1982).

Meisel, H.: Chemical characterization and opioid activity of an exorphin isolated from in vivo digests of casein. FEBS Lett. *196:* 223–227 (1986).

Mitsuma, T.; Nogimori, T.; Chaya, M.: β-Casomorphin inhibits thyrotropin secretion in rats. Exp. clin. Endocrin. *84:* 324–330 (1984).

Morley, J.E.: Food peptides – a new class of hormones? J. Am. med. Ass. *247:* 2379–2380 (1982).

Morley, J.E.; Levine, A.S.; Gosnell, B.A.; Billington, C.J.: Which opioid receptor mechanism modulates feeding? Appetite *5:* 61–68 (1984).

Morley, J.F.; Levine, A.S.; Yamada, T.; Gebhard, R.L.; Prigge, W.F.; Shafer, R.B.; Goetz, F.C.; Silvis, S.E.: Effect of exorphins on gastrointestinal function, hormonal release, and appetite. Gastroenterology *84:* 1517–1523 (1983).

Moscucci, M.: Recenti acquisizioni sulla farmacologia delle esorfine, una nuova classe di peptidi oppioidi di origine alimentare. Clin. Ter. *108:* 175–182 (1984).

Nathanson, J.A.; Glaser, G.H.: Identification of β-adrenergio-sensitive adenylate cyclase in intracranial blood vessels. Nature, Lond. *278:* 567–569 (1979).

Nedvidkovà, J.; Kasafirek, E.; Dlabac, A.; Felt, V.: Effect of β-casomorphin and its analogue on serum prolactin in the rat. Exp. clin. Endocrin. *85:* 249–252 (1985).

Neurath, H.: Proteolytic enzymes, past and present. Fed. Proc. *44:* 2907–2913 (1985).

Nieter, M.; Schatz, H.: Insulotropic action of endorphins and β-casomorphin, an opioid-like peptide fragment of milk casein. Acta endocr., Copenh. *97:* 243 (1981).

Nieter, M.; Teschemacher, H.J.; Schatz, H.: Insulotropic action of endorphins and β-casomorphine, an opioid-like peptide fragment of milk casein. Diabetologia *21:* 309 (1981).

Nilius, B.; Neubert, K.; Zett, L.; Barth, A.: β-Casomorphin affects fast and slow action potentials in ventricular myocardium. Pharmazie *40:* 428–429 (1985).

Orlowski, M.: Pituitary endopeptidases. Mol. cell. Biochem. *52:* 49–74 (1983).

Osborne, H.H.; Herz, A.: Opiate binding sites in bovine retina; evidence for benzomorphan selective binding sites. Eur. J. Pharmacol. *86:* 373–378 (1983).

Otto, D.; Karabensch, K.H.; Neubert, K.: Behaviour physiological and neurophysiological studies on the action of β-casomorphin-5 on insects. Pharmazie *37:* 793–794 (1982).

Palmieri, F.E.; Churchill, L.C.; Gerritsen, M.E.; Ward, P.E.: Enkephalin metabolism by microvascular aminopeptidase. Fed. Proc. *40:* 425 (1985).

Palmour, R.M.; Ervin, F.R.; Wagenmaker, H.; Cade, R.: Characterisation of a peptide derived from the serum of psychiatric patients. Soc. Neurosci. Abstr. *3:* 320 (1977).

Panksepp, J.; Normansell, L.; Siviy, S.; Rossi, J., III; Zolovick, A.J.: Casomorphins reduce separation distress in chicks. Peptides *5:* 829–831 (1984).

Parker, F.; Migliore-Samour, D.; Floch, F.; Zerial, A.; Werner, H.; Jolles, J.; Casaretto, M.; Zahn, H.; Jolles, P.: Immunostimulating hexapeptide from human casein: amino acid sequence, synthesis and biological properties. Eur. J. Biochem. *145:* 677–682 (1984).

Paroli, E.; Pasquarelli, V.; Pantaleoni, G.C.; Nencini, P.: Morfina, casomorfine, altri peptidi oppioidi e citoprotezione. XXIII Congr. Nazionale della Società Italiana di Farmacologia, San Remo 1986. Abstr. A, p. 89 (ECIG, Genova 1986).

Partridge, W.M.: Neuropeptides and the blood brain barrier. A. Rev. Physiol. *45:* 73–82 (1983).

Partridge, W.M.: Transport of nutrients and hormones through the blood-brain barrier. Fed. Proc. *43:* 201–204 (1984).

Partridge, W.M.; Frank, H.I.L.; Cornford, E.M.; Braun, L.D.; Crane, P.D.; Oldendorf, W.H.: Neuropeptides and the blood-brain barrier; in Martin, Riechlin, Bick, Neurosecretion and brain peptides, pp. 321–328 (Raven Press, New York 1981).

Pasquarelli, V.; Paroli, E.: Casomorfina-5 e morficeptina sulle attività enzimatiche di culture di digiuno. XXIII Congr Nazionale della Sociatà Italiana di Farmacologia, Sanremo 1986. Abstr. C, p. 89 (ECIG, Genova 1986).

Pasternak, G.W.; Wood, P.J.: Minireview: multiple mu opiate receptors. Life Sci. *38:* 1889–1898 (1986).

Peroutka, S.J.; Moskowitz, M.A.; Reinhard, J.F., Jr.; Snyder, S.H.: Neurotransmitter receptor binding in bovine cerebral micro-vessels. Science *208:* 610–612 (1980).

Petrilli, P.; Picone, D.; Caporale, C.; Addeo, F.; Auricchio, S.; Marino, G.: Does casomorphin have a functional role? FEBS Lett. *169:* 53–56 (1984).

Pfaus, J.G.; Pendleton, N.; Gorzalka, B.B.: Dual effect of morphiceptin on lordosis behavior: possible mediation by different opioid receptor subtypes. Pharmacol. Biochem. Behav. *24:* 1461–1464 (1986).

Philpott, W.H.; Khaleeluddin, K.; Philpott, K.: I. The role of addiction in the mental disease process. II. On the chemistry of addiction. J. appl. Nutr. *32:* 20–36 (1980).

Pillion, D.J.; Haskell, J.F.; Meezan, E.: Cerebral cortical microvessels: an insulin-sensitive tissue. Biochem. biophys. Res. Commun. *104:* 686–692 (1982).

Punnen, S.; Sapru, H.N.: Cardiovascular responses to medullary microinjection of opiate agonists in urethane-anesthetized rats. J. cardiovasc. Pharmacol. *8:* 950–956 (1986).

Rauca, C.; Matthies, H.: Effects of intrastriatal microinjection of β-casomorphins and its Des-tyrosine derivatives on rat motor behaviour. Biomed. Biochim. Acta *45:* 1083–1091 (1986).

Reymann, K.G.; Chepkova, A.N.; Matthies, H.: Effects of deprolorphin, a casomorphin analog, on hippocampal CA1 field potentials in vitro. Peptides *4:* 283–286 (1983).

Reymann, K.G.; Chepkova, A.N.; Schulzek, K.; Ott, T.: Effects of β-casomorphin on dentate hippocampal field potentials in freely moving rats. Biomed. Biochim. Acta *44:* 749–454 (1985).

Richardson, B.C.; Mercier, J.C.: The primary structure of the ovine β-caseins. Eur. J. Biochem. *99:* 285–297 (1979).

Richter, W.O.; Schwandt, P.: Peptide hormones and lypolisis in rabbit adipocites. Hormone metabol. Res. *17:* 127–130 (1985).

Rogers, S.; Wells, R.; Reichsteiner, M.: Amino acid sequences common to rapidly degraded proteins: the PEST hypothesis. Science *234:* 364–368 (1986).

Sander, G.E.; Giles, T.D.: Enkephalin analogs and dermorphin in the conscious dog: structure-activity relationships. Peptides *3:* 1017–1021 (1982).

Sapru, H.N.; Willtte, R.N.; Krieger, A.J.: Stimulation of pulmonary J receptors by an enkephalin analog. J. Pharmac. exp. Ther. *217:* 228–234 (1981).

Schaeffer, J.I.; Haddad, G.G.: Regulation of ventilation and oxygen consumption by δ- and μ-opioid receptor agonists. J. appl. Physiol. *59:* 959–968 (1985).

Schick, R.; Holland, A.; De La Fuente, A.; et al.: Effect of milk-derived opiates on pancreatic polypeptide secretion in dogs. Acta endocr., Copenh. *102:* 80–81 (1983).

Schick, R.; Schusdziarra, V.; Brantl, V.; De La Fuente, A.: Effect of β-casomorphine on insulin release on dogs. Acta endocr., Copenh. *99:* suppl. 246 (1982).

Schick, R.; Schusdziarra, V.; Rewes, B.; Holland, A.; Lenz, N.; Pfeiffer, E.F.: Effect of exogenous and endogenous opiates on gastrin release in dogs. Dig. Dis. Sci. *29:* 75S (1984).

Schusdziarra, V.; Henrichs, I.; Holland, A.; Klier, M.; Pfeiffer, E.F.: Evidence for an effect of exorphins on plasma insulin and glucagon levels in dogs. Diabetes *30:* 362–364 (1981).

Schusdziarra, V.; Holland, A.; Schick, R.: Modulation of post-prandial insulin release by ingested opiate-like substances in dogs. Diabetologia *24:* 113–116 (1983a).

Schusdziarra, V.; Schick, R.; De La Fuente A.; Holland, A.; Brantl, V.; Pfeiffer, E.F.: Effect of β-casomorphins on somatostatin release in dogs. Endocrinology *112:* 1948–1951 (1983b).

Schusdziarra, V.; Specht, J.; Schick, R.; De La Fuente, A.; Holland, A.; Pfeiffer, E.F.: Effect of morphine, Leu-enkephalin and β-casomorphins on basal somatostatin release. Hormone metabol. Res. *15:* 407–408 (1983c).

Schusdziarra, V.; Schick, A.; De La Fuente, A.; Specht, J.; Klier, M.; Brantl, V.; Pfeiffer, E.F.: Effect of β-casomorphins and analogs on insulin release in dogs. Endocrinology *112:* 885–889 (1983d).

Schusdziarra, V.; Schick, R.; Holland, A.; De La Fuente, A.; et al.: Effect of opiate-active substances on pancreatic polypeptide levels in dogs. Peptides *4:* 205–210 (1983e).

Sharma, S.K.; Nirenberg, M.; Klee, W.A.: Morphine receptors as regulators of adenylate cyclase activity. Proc. natn. Acad. Sci. USA *72:* 590–594 (1975).

Shinoda, I.; Tada, M.; Okai, H.; Fukui, S.: Bitter taste of H-Pro-Phe-Pro-Gly-Pro-Ile-Pro-OH corresponding to the partial sequence (position of 61–67) of bovine β-casein and related peptides. Agric. Biol. Chem. *50:* 1247–1254 (1986).

Shook, J.E.; Burks, T.F.: Peripheral mu-opioid receptors mediate small intestinal transit (SIT) bot not analgesia in mice. Fed. Proc. *45:* 1344 (1986).

Silk, D.B.A.: Progress report. Peptide absorption in man. Gut *15:* 494–501 (1974).

Simantov, R.; Lotem, J.; Levy, R.: Selectivity in the control of opiate receptor density in the animal and in the cultured fetal brain cells. Neuropeptides *5:* 197–200 (1984).

Singh, M.M.; Kay, S.R.: Wheat gluten as a pathogenic factor in schizophrenia. Science *191:* 401–402 (1976).

Smagin, V.G.; Titov, M.I.; Shatalov, V.N.; Vinogradov, V.A.; et al.: Capacity of opiate receptor ligands (endorphins and exorphins) to inhibit gastric juice secretion in dogs. Byull. Eksp. Biol. Med. *95:* 78–80 (1983).

Stan, E.Y.; Aleinik, S.I.; Chernikov, M.P.: Physiologically active κ-casein peptides. Fiziol. Zh. SSSR im I.M. Sechenova *69:* 885 (1983).

Stark, H.; Lossner, B.; Matthies, H.: Degradation of β-casomorphins in the rat brain in vivo. Biomed. Biochim. Acta *45:* 557–563 (1986).

Svedberg, J.; De Haas, J.; Leimenstoll, G.; Paul, F.; Teschemaker, H.: Demonstration of β-casomorphin immunoreactive materials in vitro digests of bovine milk and in small intestine contents after bovine milk ingestion in adult humans. Peptides *6:* 825–830 (1985).

Teschemacher, H.; Ahnert, G.; Umbach, M.; Kielwein, G.; Leib, S.: β-Casomorphins-opiate-like acting peptide fragments from β-casein: determination in various milk and tissue extracts by radioimmunoassay. Arch. Pharmacol. *311:* R67 (1980).

Teschemacher, H.; Umbach, M.; Hamel, U.; Praetorius, K.; Ahnert-Hilger, G.; Brantl, V.; Lottspeich, F.; Henschen, A.: No evidence for the presence of β-casomorphins in human plasma after ingestion of cow's milk or milk products. J. Dairy Res. *53:* 135–138 (1986).

Traber, J.; Dompert, W.; Schollnhammer, G.; Gispen, W.H.: Opiate-like and neuroleptic-like activities of β-casomorphins in rat. Neurosci. Lett. *18:* suppl., p. S378 (1984).

Umbach, M.; Teschemacher, H.; Praetorius, K.; Hirschauser, R.; Bostedt, H.: Demonstration of a β-casomorphin immunoreactive material in the plasma of newborn calves after milk intake. Regulat. Peptides *12:* 223–230 (1985).

Verebey, K.; Volavka, J.; Clouet, D.: Endophins in psychiatry. Archs. gen. Psychiat. *35:* 877–888 (1978).

Villiger, J.W.: Morphiceptin: biphasic Inhibition of μ-opiate receptor ligands. Pharmacol Biochem. Behav. *21:* 705–706 (1984).

Wajda, I.J.; Neidle, A.; Ehrenpreis, S.; Manigault, I.: Properties and distribution of morphine-like substances; in Kosterlitz, Opiates and opioid peptides, pp. 129–136 (North-Holland/Elsevier, Amsterdam 1976).

Walter, R.; Shlank, H.; Glass, J.D.; Schwartz, I.L.; Kerenyi, T.D.: Leucylglycinamide released from oxytocin by human uterine enzyme. Science *173:* 827–828 (1971).

Watson, S.; Berger, P.; Akil, H.; Mills, M.; Barchas, J.: Effects of naloxone on schizophrenia: reduction in hallucinations in a subpopulation of subjects. Science *201:* 73–76 (1978).

Wei, E.T.; Lee, A.; Chang, J.K.: Cardiovascular effects of peptides related to the enkephalins and β-casomorphin. Life Sci. *26:* 1517–1522 (1980).

Werz, M.A.; Macdonald, R.L.: Opioid peptides with differential affinity for mu and delta receptors decrease sensory neuron calcium-dependent action potentials. J. Pharmac. exp. Ther. *227:* 394–402 (1983).

Werz, M.A.; Macdonald, R.L.: Dynorphin and neoendorphin peptides decrease dorsal root ganglion neuron calcium-dependent action potential duration. J. Pharmac. exp. Ther. *234:* 49–56 (1985).

Widy-Tyszkiewicz, E.; Czlonkowski, A.; Szreniawski, Z.: Cardiovascular responses to morphiceptin in spontaneously hypertensive and normotensive rats. Pol. J. Pharmacol. Pharm. *38:* 51–56 (1986).

Wienbeck, M.; Bielfeld, P.; Karaus, M.; Erckenbrecht, J.F.: Is gut motility affected by exorphins? Dig. Dis. Sci. *30:* 800 (1985).

Williams, C.L.; Heyman, J.S.; Porreca, F.; Burks, T.F.: Cerebral, spinal and peripheral inhibition of gastrointestinal transit by PLO17: differential antagonism by naloxonazine. Fed. Proc. *45:* 1053 (1986).

Wohltmann, M.; Roth, B.L.; Coscia, C.J.: Differential postnatal development of mu and delta receptors. Dev. Brain Res. *3:* 679–684 (1982).

Yonehara, N.; Clouet, D.H.: Effects of delta and mu opiopeptides on the turnover and release of dopamine in rat striatum. J. Pharmac. exp. Ther. *231:* 38–42 (1984).

Yoshikawa, M.; Tani, F.; Yoshimura, T.; Chiba, H.: Opioid peptides from milk proteins. Agric. Biol. Chem. *50:* 2419–2421 (1986).

Yoshikawa, M.; Yoshimura, T.; Chiba, H.: Opioid peptides from human β-casein. Agric. Biol. Chem. *48:* 3185–3187 (1984).

Yu, V.C.; Richards, M.L.; Sadèe, W.: A human neuroblastoma cell line expresses μ and δ opioid receptor. J. biol. Chem. *261:* 1065–1070 (1986).

Zadina, J.E.; Kastin, A.J.: Interaction between the antiopiate Tyr-MIF-1 and the mu opiate morphiceptin at their receptive binding sites in the brain. Peptides *6:* 695–970 (1985).

Zhang, A.Z.; Chang, J.K.; Pasternak, G.W.: The actions of naloxazone on the binding and analgesic properties of morphiceptin (NH_2-Tyr-Pro-Phe-Pro-$CONH_2$), a selective mu-receptor ligand. Life Sci. *28:* 2829–2836 (1981).

Zioudrou, C.; Klee, W.A.: Exorphins – peptides with opioid activity derived from alpha-casein and wheat gluten; in Van Ree, Terenius, Characteristics and functions of opioids, pp. 243–244 (Elsevier/North-Holland Biomedical Press, Amsterdam 1978).

Zioudrou, C.; Klee, W.A.: Possible role of peptides derived from food proteins in brain function; in Wurtman, Wurtman, Nutrition and the brain, vol. 4 (Raven Press, New York 1979).

Zioudrou, C.; Streaty, R.A.; Klee, W.A.: Opioid peptides derived from food proteins. J. biol. Chem. *254:* 2446–2449 (1979).

Prof. Eugenio Paroli, Institute of Medical Pharmacology and Drug Addiction Center, University of Rome 'La Sapienza', Aldo Moro 5, I–00186 Rome (Italy)

Wld Rev. Nutr. Diet., vol. 55, pp. 98–152 (Karger, Basel 1988)

Selenium-Related Endemic Diseases and the Daily Selenium Requirement of Humans

Guangqi Yang, Keyou Ge, Junshi Chen, Xiaoshu Chen

Institute of Nutrition and Food Hygiene, Chinese Academy of Preventive Medicine, Beijing, China

Contents

I. Introduction

Selenium was first recognized as a food toxic hazard in the seleniferous area of the USA in the 1930s [Franke, 1934]. Twenty years later, Schwarz and Foltz [1957] discovered, however, that minute quantities of selenium in the diet protected against nutritional liver necrosis in vitamin E-deficient rats. Following the production of certain uncomplicated selenium deficiency diseases in the laboratory animals [Scott et al., 1967; McCoy and Weswig, 1969; Thompson and Scott, 1970; Wu et al., 1973], selenium was evidenced to be an essential nutrient for animals. In China, regional disease characteristics of biogeochemical origin, such as goiter, fluorosis, selenosis, Keshan disease (KD) and Kashin-Beck disease (KBD), etc., are put into the class showing endemic disease. As early as 2,000 years ago, the close relationship between human health and its environment had been stated in the famous traditional Chinese medical writings, Huang De Nei Jin Su-wen [Zhu, 1983]. Billions of inhabitants living in the endemic area were afflicted by the menace of the diseases. Conclusive evidence showed that selenium was implicated in human diseases in relation to either excessive or insufficient intake was first reported in 1966 in the former case [Yang et al., 1981] and in 1974 in the latter case [Keshan Disease Research Group of the Chinese Academy of Medical Sciences, 1979a; Chen et al., 1980].

The association of cancer with low selenium suggested that selenium status might be able to modify the risk of cancer and thus has received great attention in recent years [Schrauzer et al., 1977]. An inverse correlation between liver cancer mortality and selenium levels in blood and grains in an area with high risk of hepatoma has also been reported in this country [Yu et al., 1985]. Much work has also been done on human selenium requirement, since the existence of extremely low and high selenium areas in China, it provided a unique place for this study [Yang, 1985a, b].

The fact that biogeochemical disease usually appeared among populations subsisting largely on the farm products produced from local soil was clearly demonstrated in China. The traditional Chinese rural diet mainly consists of locally produced vegetable components. It affords the imbalance state of selenium in the local soil to be transferred through the diet to the inhabitants and brought about related diseases.

II. Environment Selenium Levels in Endemic Areas and the Selenium Intakes of Inhabitants

Selenium-related diseases were first recognized by the local inhabitants as a 'water and soil' disease, since they believed that these diseases must be caused by something in the local water or soil. In seleniferous area, residents claimed that drinking water effluent from the coal bed was adverse to the health. This view led the later discovery of an extraordinary selenium-rich stony coal [Yang et al., 1983]. Both KD and KBD distributed at eroded areas such as the mountain slopes and hillsides. Inhabitants in endemic areas experienced that one could escape from KD if one moved frequently from an affected area to a non-affected area. The concomitant occurrence of white muscle disease in domestic animals in the endemic areas gave the clue that selenium may be deficient in the environment. In 1973, it was first revealed that the selenium status of KD patients is very poor, for instance, the average blood and hair selenium concentration of 40 KD patients were as low as only 0.018 and 0.107 ppm, respectively [Yang and Wang, 1973, unpublished]. This result was later confirmed by many workers in China [Tan, 1982; Xu, 1982; Yang and Du, 1983]. Selenium estimations in different materials were carried out by a modification of Watkinson's fluorometric method [Watkinson, 1966] in our laboratory [Wang et al., 1979, 1985; Liu et al., 1985; Sun and Yang, 1985].

A. Selenium Concentration of Soil, Drinking Water and Foods

The concentration of selenium in foods of plant origin in particular depends largely on the amounts available in the soil. Where there is notable insufficiency or excess in the soil, selenium deficiency or excess diseases may become endemic through the food chain. Extraordinarily high and low concentrations of selenium had been detected in the environment of endemic areas. Data obtained from analysis of soil, water and staple grains from different parts of China are compiled in table I [Yang et al., 1983; Sun and Yang, 1985; Sun et al., 1985; Yang, 1985b]. It was reported

Table I. Selenium concentration of soil, water and staple grains in different areas of China

Items	Selenium content, μg/g		
	low	medium	high
Soil[1]	0.112±0.024 (44)	0.234±0.035 (31)	(6.39–10.66)[3]
Water	trace−0.0004 (8)	trace−0.005 (31)	0.054 (11) (0.009–0.159)[3]
Corn[2]	0.005±0.003 (195)	0.036±0.056 (82)	8.1 (44) (0.5–28,5)[3]
Rice[2]	0.007±0.003 (49)	0.035±0.027 (76)	4.0 (22) (0.3–20.2)[3]
Soybean[2]	0.010±0.008 (150)	0.069±0.076 (31)	11.9 (17) (5.0–22.2)[3]

Numbers in parentheses are the number of samples analyzed.
[1] Mean ± SEM.
[2] Mean ± SD.
[3] Range.

that the ordinary selenium content of soil lies between 0.1 and 2 ppm [NRC, 1983]. However, in our low selenium area, the soil selenium content, 0.112 μg/g (table I), is near the lower extremity and is lower than the Finnish agricultural soil concentration of 0.2–0.3 μg/g [Koivistoninen, 1986]. Furthermore, many factors, such as the lower pH, the poor drainage or the higher humus and iron content in the local soil may decrease the bioavailability of selenium in soil in a poor selenium area. The amount of water-soluble selenium in the soil of a low selenium area was measured only 3.9% of the total which is significantly less than the 9.2% measured in selenium-adequate area. In the high selenium area, the soil selenium concentration rose to 6–11 ppm, which reached 30- to 50-fold the ordinary selenium content (table I). Since the soil was made alkali by the use of lime and plant ash as fertilizers in the high selenium area, the selenium in drinking water was almost exclusively oxidized to the form of selenate which is highly available to the plant [Sun et al., 1985].

Drinking water usually contained a negligible amount of selenium in respect to the daily selenium intake. It is probably true both in selenium-deficient and selenium-adequate areas. In a seleniferous area, although it contained considerable amounts of selenium in the water, it is still far less important a source than the dietary staple cereals. For example, the average selenium concentration of 44 corn samples averaged 8.1 μg/g in contrast to the average value of 0.054 μg/ml obtained from 11 water samples [Yang et al., 1983]. Morris and Levander [1970] reported from the USA that the selenium concentration of soybean from the US market averaged 0.070 μg/g which is comparable to the selenium concentration of water in our seleniferous area, but is still far less than an average of 11.9 μg/g of the same food estimated in this area. It seems that although selenium in water always presents in trace amounts or some samples even contain undetectable amount of selenium, it yet serves to reflect the specific geochemical status to some extent in different environments.

B. Daily Selenium-Intake in Different Parts of China

The selenium intake of the Chinese rural population is derived largely from grains. Because of the great variety in selenium content of grains from different areas, the dietary selenium intake would be expected to vary greatly. The variation in daily selenium intakes in different areas of China is shown in table II.

Table II. The daily selenium intakes in selenium deficiency, excess and adequate areas of China in comparison with that of other countries

Country	Dietary selenium intake, µg/day	Reference
China (KD-affected area)	7–11	Yang [1985a]
China (near KD-affected area)	17	Yang [1985a]
Finland	20–30	Koivistoinen [1986]
China (Beijing)	66	Yin et al. [1985a]
USA (market basket)	169	Levander [1983]
Venezuela (seleniferous area)	300–400	Brätter et al. [1984]
China (near by selenosis area)	750	Yang [1985a]
China (chronic selenosis area)	4,990	Yang [1985a]

C. The Influence of Dietary Habit on Selenium Intake

Although an unfavorable geochemical environment has a profound influence on the health of humans, dietary habit can aggravate or alleviate, or even eliminate the hazardous effect. It was illustrated by a comparison of the selenium intakes of Chinese and New Zealanders with their dietary habit [Yang, 1985b]. As the selenium contents of the main foods and the selenium intakes by the residents of both countries are compared, it was found that the differences in the selenium content of foods are not so great as to account for the rather large difference in daily selenium intakes (table III). However, this could well be explained by the pronounced difference in their dietary habit. New Zealanders obtained 82% of their selenium intakes from animal foods which have a rather higher selenium concentration than plant sources in the similar low selenium area; while Chinese derived 75% of their selenium from cereals produced in local soil, and animal sources only accounted for 6.9%. Obviously even in the same low selenium environment, the major dietary selenium source of the Chinese residents should provide less selenium than that of the New Zealanders. Selenium status of the Chinese population in endemic areas were only a half to a third the levels of those of the latter. The results are shown in table IV. The average blood selenium level of 59 ng/ml in New Zealand residents is comparable to the average level of 52 ng/ml in inhabitants of the KD area after receiving selenite supplements for 6 months, and this treatment was shown to be effective in protecting against KD in the pre-

Table III. Comparison of selenium content of foods and dietary selenium intakes in low selenium areas in China and New Zealand. From Yang [1985b]

County	Food selenium μg/g			Dietary selenium intake μg/g	Sources, %		
	cereals	cabbage	egg		cereal	animal foods	drinking water
New Zealand	0.011	0.0008	0.24	6–70	14–18	82	negligible
China	0.006	0.0007	0.12	4–11	75	7	negligible

Table IV. Selenium status of residents in low selenium areas in China and New Zealand

Country	Blood ng/ml	Plasma ng/ml	Red cell ng/ml	Toe-nail μg/g	Milk ng/ml	Urine μg/day	References
New Zealand	59	48	74	0.26	7.6 13.1	12.7	Robinson [1984][1]
China	18	22[2]	31[2]	0.17	2.6	4.8	Yang [1985b]

[1] Personal communication.
[2] Xi [1985, personal communication].

Table V. Selenium status of inhabitants in different parts of China

Areas of different levels of selenium	Hair, ppm	Whole blood, ppm	Urine, μg/ml
High	32.2 (65) (4.1–1.00)[1]	3.2 (72) (1.3–7.5)[1]	2.68 (17) (0.88–6.63)[1]
Medium	0.36 ± 0.17 (1,745)	0.095 ± 0.091 (111)	0.026 ± 0.012 (19)
Low	0.074 ± 0.050 (1,478)	0.018 ± 0.010 (325)	0.007 ± 0.001 (43)

Numbers in parentheses are the number of samples analyzed. Results are given as mean ± SD.
[1] Range.

vious trial [Chen et al., 1980]. It suggests that KD appears only under extremely poor selenium status. Furthermore, animal foods were known to be able to protect residents against KD, since herdsman usually have a higher proportion of animal foods and never suffered KD in the endemic areas, even though their average hair selenium concentration was at the marginal level of deficiency. The exact reason is still obscure; perhaps other unknown factors are involved. The above-mentioned example clearly demonstrated the fact that dietary habit plays a very important role in protecting the inhabitants against the adverse effects of environment.

D. Selenium Status of Inhabitants in Different Parts of China

Since there is a wide range of selenium intakes in China, a wide range of selenium levels in human tissues, blood and urine is expected and the data are shown in table V. In view of the striking differences between the tissue selenium levels in the high and low areas one will not be surprised to see the occurrence of endemic diseases in these areas.

III. Endemic Selenosis

A. History

Marco Polo first described an animal disease in 1295 in Juquan county of Gansu province in his travel to the northwest of China [Marco Polo, 1937]. The symptoms are similar to those of the animal selenosis. Subsequent reports in China did not appear until 1966 when an endemic human disease of unknown cause characterized by loss of hair and nails heavily prevailed. This disease occurred long ago in Enshi County of Hubei Province, in the southwest of China. It broke out at irregular intervals. The recent outbreak was from 1961 to 1964. The average incidence rate in five heavily prevalent villages was 49.2% in 248 inhabitants and reached 82.5% among 22 inhabitants in the most severely affected village. The cases were diagnosed by pronounced loss of hair or diseased nails or both. Domestic animals and certain species of insects and crops had also suffered severely. An extremely high selenium concentration was subsequently detected in the stony coal bed [Yang et al., 1981]. The focal distribution of seleniferous areas in this region obviously followed the distribution pattern of the coal stratum, and the exact extent of this seleniferous region, however, still remains unknown.

B. Identification of the Cause

1. Animal Experiment

(1) Signs and Growth Response

The locally 'toxic' corn was diluted with an equal weight of marketed corn from Beijing and the mixed flour was used to feed rats. Twenty-two weanling rats were randomly divided into two groups of equal numbers. All the control animals receiving the mixed flour diet died after 32 days on diet. The selenium content of the toxic corn was estimated to be 44 μg/g. Since 5 μg/g arsenic in drinking water was reported to alleviate selenium toxicity in animals [Moxon, 1938], drinking water containing arsenic from arsenic trioxide in the range of 5–15 μg/g was provided for the animals in the experimental group. Results showed that the toxicity of the grain was successfully alleviated by arsenic, but recovered when arsenic was withdrawn from the drinking water as judged by the corresponding changes in body weight. Most of our animals intoxicated also showed severe anemia. Average hemoglobin contents dropped from 11.9 (8.3–13.8) to 8.2 (3.8–13.4) g/100 ml on the day before death [Yang et al., 1981].

(2) Pathological Change of Selenium Toxication

All animals in the experimental group show severe jaundice and distinct golden-yellow ring appeared around the opening of the genitourinary organ. Livers of the experimental animals were severely damaged and large area of hemolysis could be observed by the naked eye. The pathologic changes in selenium-poisoned animals were summarized by Rosenfeed and Beath [1946]. Morphological changes of livestock intoxicated with various organic and inorganic selenium compounds were properly described in this book. Studies on selenium poisoning in rats and dogs have also been carried out in our laboratory [Zhang et al., 1974]. Intravenous injection of sodium selenite 0.1 mg/kg body weight daily induced liver edema and congestion in dogs. The parenchymal cells are swollen and granular. When the dosage was increased to 1.0 mg/kg, extensive necrosis occurred with distinct confluent hemorrhage throughout the liver. The lobule periphery zone and portal area are most severely affected (fig. 1).

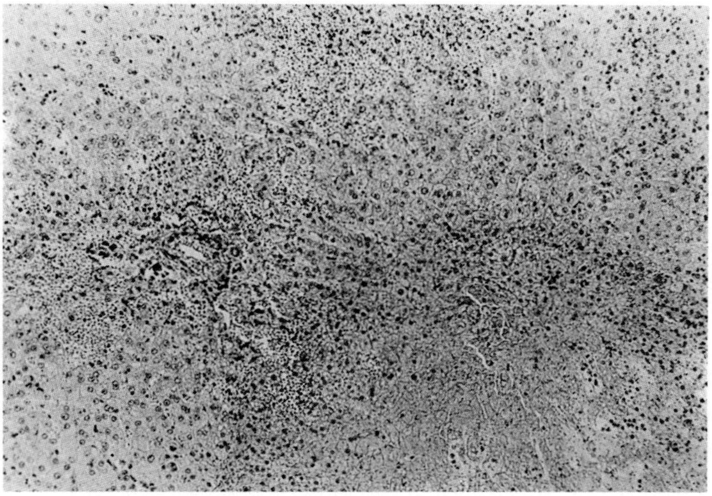

Fig. 1. Coagulation necrosis of liver of dog in selenium intoxication. HE. × 120.

The principal injury of the kidney occurs in the convoluted tubules. Hemorrhages were seen in heart, lungs, spleen and many other organs; encephaledema may occur in some cases.

2. Field Study

The geographical distribution of this disease depends upon whether the selenium-rich coalmine is deposited or not. All affected areas have their coalmines exposed. The average selenium content of 10 coal samples was estimated to be 291 μg/g ranging from 128 to 834; one sample of 3,632 μg/g was excluded. When another 9 samples were collected and estimated, the average was 367 μg/g and a tenth sample assayed as 84,123 μg/g. In a joint China-USA collaborative selenium analysis, the tenth sample was sent to Prof. O.E. Olson's laboratory through the help of Dr. O.A. Levander in Beltsville and assayed as 92,800 μg/g. Weathering, leaching or biological action would gradually mobilize selenium from the selenium-rich coal to the soil and thus make it available to the crops. The traditional use of coal ash and lime as fertilizers would not only render the selenium accumulated in the cultivated soil but also make it more readily available to crops. Moreover, the unusually high selenium content of the coal might heavily contaminate the air and foodstuffs by its smoke and contribute additionally to the high selenium intake of the residents in the site where the disease prevailed (table II). Additional factors were believed necessary to account for this particular outbreak of selenosis, e.g. the drought led to a failure of the rice crop which forced the villagers to consume more high selenium vegetables and maize instead. Furthermore, the residents consumed fewer protein foods and would likely be more sensitive to the toxic action of excessive selenium intake [Smith, 1939].

C. Signs of Selenosis in Humans [Yang et al., 1983]

The tissues mostly affected were hair, finger and toenail, skin, possibly teeth, and the nervous system. For instance, among the 18 cases in the heavily affected village, 100% have total or partial hair loss and nails dropping, 28% have skin lesions.

1. Loss of Hair and Nails

The hair becomes dry and brittle and is easily broken off at the scalp. The appearance of a rash on the scalp is associated with intolerable itching. Usually hair can be removed by hard scratching. When hair is broken off, the radicles remain intact so that the hair continues to grow. New hair is always depigmented and looses its luster; sometimes the ends are forked.

White areas and longitudinal streaks usually appear on the nail surface. The nails become brittle and thick, and this is followed by a break on the wall of the nail. As new growth continues, the broken nail is pushed forward and finally drops off. In many cases, fluid effuses from the skin around the nail. In those cases, a much longer period is needed to finish the whole process. The new nail is deformed (fig. 2). However, both hair and nails can completely recover to normal after changing the diet for a certain period of time.

2. Skin Lesions

Affected skin becomes red and swollen and then blistered and eruptive. In some cases ulcerations follow that take a long time to heal. Reddish pigmentation of the skin usually remains.

3. Tooth Decay

Nearly one third of 46 cases had heavily mottled tooth enamel and in some of them the teeth were eroded or had pits. The appearance of mottled enamel tooth in this area was believed to be related to fluorosis. No evidence has yet been obtained that excessive exposure to selenium contrib-

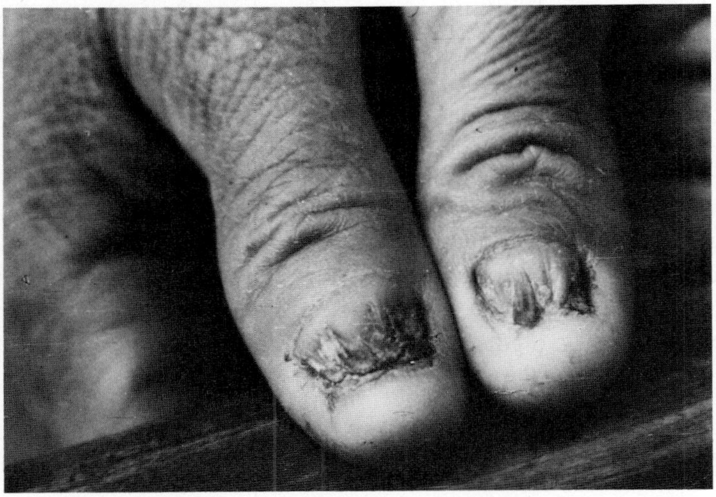

Fig. 2. A case of nail loss showing deformation of newly grown nail.

utes to the severity of such a type of tooth damage, although it has been reported that excessive exposure to selenium increases dental caries in man and animals [Hadjimarkos, 1973].

4. Abnormalities of the Nervous System

Pronounced abnormalities of the nervous system were observed only in one heavily affected village. Initially the patients may complain of peripheral anesthesia and pain in the extremities; then numbness, convulsions, paralysis, motor disturbance and even hemiplegia may develop. It is likely that all clinical neurological signs are due to polyneuritis caused by intoxication. Among the 18 patients in this village, 3 had paresthesia and one with hemiplegia died. Fluorosis in this area may contribute to this toxic effect particularly in mild cases. Disturbances of the digestive tract usually accompanied this type of intoxication.

D. Selenosis of Livestock

Alkali diseases of cattles, buffaloes, and pigs were not uncommon in the affected areas. Water buffalo seemed more susceptible than those of cattle. In addition to the damage of hoofs, the horns are also sloughed. It appeared that poor hatchability and failing to raise piglets are the most sensitive parameters indicating the appearance of selenosis in the localities.

E. Damage to Plants

Although it was suggested previously that plant injury due to selenium overexposure would probably not occur under natural conditions, injuries to crops did occur in this area. 'White seedlings' of corn plant occurred more frequently and leaf edges of injured plants usually became white. The corn kernels stuck firmly to the cob and matured at a smaller size.

F. Prevention

Since the source of selenium is coal, if the coal is worthy of mining, the problem may gradually be solved; if not, the following steps have to be taken:

(1) Selection of crops for planting. It is necessary to get a clear understanding of the distribution of the coal bed and its potential to contaminate the soil in different localities of this seleniferous area. In the heavily contaminated sites only economic crops such as cotton, sugar or oil crops are encouraged. Afforestation is also a promising way for reasonable utilization of the high selenium land.

(2) Prevention of the soil from contamination with coal and biological avail of its selenium with ease. Coal ash used as fertilizer should be prohibited and application of lime should also be under control.

(3) Proper dilution of seleniferous grains with products from normal areas.

(4) Transportation of seleniferous staple cereal to the low selenium areas. The shortest distance from the site of high selenium to low selenium KD area, for instance, is less than 100 km. Transportation or exchange of staple foods among areas would afford a practical measure for the control of both endemic diseases.

(5) Change of dietary habit. Appropriate increase of animal foods from the traditional vegetable diet would effectively decrease the amount of selenium intake.

Although it was proposed that addition of sulfur or gypsum to the seleniferous soil might reduce selenium uptake by plants [Underwood,

1977], it is doubtful whether this method can be applied in this area with success, since there is already an abundance of sulfur in the soil originally from an iron sulfide mine in this seleniferous area.

IV. Selenium Deficiency and Endemic Diseases

A. Keshan Disease

1. History

No reliable record is available for tracing the history of when KD made its first appearance. According to the ancient stele and county records it is very likely that KD first occurred 100 years ago in northeast and northwest China. It used to break out in these districts in the uncultivated land where migrants from other parts of China gathered. Attention was attracted only during its outbreak in 1935 in Keshan County and its neighboring counties. It showed high incidence and fatality, and was erroneously suspected as an infectious disease. According to the survey on 7 counties in that year, young and middle-aged women were most susceptible. The common signs of this unknown disease were precardial oppression and pain, nausea and vomiting of yellowish fluid. The intervals from getting sick until fatal termination were within one or two days in most cases. This unknown disease was later named Keshan disease. In the 1960s this disease was also discovered in southwestern China. Early in 1962, Zheng emphasized the similarities between lamb white muscle disease and human KD in a national congress of animal husbandry in China [Zheng, 1962, unpublished]. Small-scale human selenium intervention studies were carried out in 1965 by Xian Medical College, and by the Keshan Disease Research Group of the Chinese Academy of Medical Sciences in 1969. No conclusive evidence could be drawn as to the effect of supplementation either with selenium in combination with vitamin E [Farm Hospital of Fu county and Xian Medical College, Shanxi, unpublished] or with selenium alone [Keshan Disease Research Group, unpublished], although supplementation invariably seemed favorable in each trial. Extremely low selenium levels in blood and hair of patients with KD and residents in the KD area were first reported in 1973 in our laboratory [Yang et al., 1973, unpublished]. Conclusive evidence of its effectiveness in the prevention of KD was obtained in a heavy prevalence area in the southwest of China in 1974 by Keshan Disease Research Group of the

Chinese Academy of Medical Sciences [1979a, b]. The same result was reached by Xian Medical College [1979] in 1975 and others in the years following.

2. Epidemiology of KD and Selenium

(1) Epidemiological Characteristics of KD and Selenium Status

Earlier epidemiologic studies revealed that inhabitants in the northeast of China subsisting largely on rice were protected against KD. It was suggested that the causative agent is transmissive by corn or wheat but not by rice. As KD was later discovered in the southwest of China where residents mainly consumed rice, it caused further confusion. Data from our laboratory in 1974 [Yang et al., 1980] indicated that the selenium concentration of rice from the southwest of China (30 samples averaged 0.007 ppm) was significantly lower than from the northeast (25 samples averaged 0.024 ppm), while selenium in corn and wheat from the northeast (41 and 11 samples averaged 0.003 and 0.006 ppm, respectively) were just as low as those of rice from the southwest. Hair selenium levels of a rice-consuming population in the northeast of China were found to be correspondingly higher. It clearly demonstrated that the risk of KD appeared only in populations with the poorest selenium intake even in the same endemic area. Besides population susceptibility, the most distinctive epidemiological characteristics of KD are regional distribution and seasonal prevalence. It is of special interest to further study these characteristics in connection with the selenium status.

Regional Distribution. The distribution of KD shows a belt-like region throughout the mainland of the country from northeast to southwest with specific focal distribution within a wide belt and it was demonstrated that this belt is inconsistent with the area of selenium deficiency [Tan, 1982]. Affected and non-affected areas within the belt are relatively stable. The chief characteristics of the focal distribution are: KD usually occurred in hilly and mountainous areas, but not in plains or coastal areas nearby. The affected localities are usually of lower soil salinization. There are 'safety islands' with well-defined boundary lines within an affected area. Selenium levels in hair and staple food samples collected in these different regions agree reasonably well with the characteristic distribution of KD, i.e. the selenium status of all affected areas are invariably poorer than that of the nearby non-affected areas. Of particular interest is the fact that the distribution of KD in Lichuan County, a county near to the high selenium area, again shows that KD appeared restrictedly in the area, apart form a high selenium area, where the selenium levels in soil and grains are just as low as other affected areas [Yang et al., 1982a, 1984c]. As an endemic disease, this feature of distribution is of special significance since the appearance of KD is epidemiologically decided exactly by the environmental selenium status.

Seasonal Prevalence. The peak seasonal prevalence of KD in North China is in winter, but in South China it is in summer. Cases appearing in peak seasons in Sichuan Province accounted for 70–80% of the total patients in a year [Chen, 1982]. It suggests that seasonal fluctuations in body selenium levels may occur, if selenium deficiency alone is responsible for the occurrence of this disease. A total of six affected sites in Heilongjiang, Shandong, Sichuan and Yunnan provinces were selected for this study. Hair selenium concentrations of local children were monitored in four seasons in as short as one year to as long as three consecutive years. It was found that seasonal variations of hair selenium level were within the usual deficiency range; no correlation was found between the peak seasonal prevalence and levels of hair selenium fluctuation [Sun et al., 1980].

(2) Selenium Levels of Staple Foods and Human Tissues in Endemic Areas [Yang et al., 1982, 1984c]

Selenium Concentration of Staple Cereals. All staple cereal samples from affected areas contained selenium not >0.02 ppm while samples containing selenium >0.02 ppm were exclusively from non-affected areas.

Selenium Concentration and Glutathione Peroxidase (GSHPx) Activity in the Blood. The whole blood selenium content of 325 subjects living in an endemic area averaged 0.018 ± (SE) 0.001 ppm and that of 134 subjects living in the non-endemic area was 0.093 ± (SE) 0.008 ppm. The differences were remarkable. Whole blood GSHPx activities in children in the endemic area were significantly lower than those in the non-endemic areas. Again no significant differences were found between affected children and normals in the same affected area [Zhu et al., 1982] as reported for selenium previously [Yang et al., 1982]. A modification of Hafeman's method [Hafeman et al., 1974] was used for the estimation of GSHPx activity.

Selenium Concentration of Hair. The selenium content of hair is significantly correlated with blood selenium concentration ($r = 0.82$, $p < 0.001$) in ranges from 0.002 to 0.266 ppm for blood and 0.023 to 0.890 ppm for hair [Wang et al., 1979]. The average hair selenium concentration of populations in endemic areas were below 0.123 ppm and those

Table VI. Selenium concentration of human tissues[1] of patients suffering from Keshan disease (KD) or from hospitals. From Zhou and Yang [1986, unpubl.]

Patients	Selenium content, µg/g			
	skeleton muscle	heart	liver	kidney
Fresh cases of KD	0.040 (2) (0.039–0.041)[2]	0.077 (9) (0.026–0.129)[2]	0.094 (4) (0.047–0.149)[2]	0.254 (4) (0.184–0.388)[2]
Patients from hospital in an endemic area	0.074 (4) (0.043–0.105)[2]	0.067 (5) (0.026–0.141)[2]	0.075 (5) (0.043–0.120)[2]	0.173 (5) (0.045–0.405)[2]
Patients from hospital in a non-endemic area	–	0.976 (3) (0.914–1.060)[2]	1.202 (3) (1.108–1.326)[2]	2.211 (3) (2.002–3.058)[2]

Numbers in parentheses are the number of samples analyzed.
[1] Fresh weight.
[2] Range.
Part of the samples were kindly provided by Dr. Lius.

near the endemic areas ranged between 0.123 and 0.165 ppm. In non-endemic areas far apart from endemic areas it was always higher than 0.2 ppm. When hair selenium levels were expressed in county averages and were arranged in decreasing order, it was found that all affected counties were located at the lower extremity with only two exceptions, it indicated that affected areas are consistantly in selenium insufficiency, but the reverse is not necessarily true. Factor(s) other than selenium may be involved.

Selenium Concentration of Human Tissues. Selenium concentration of skeleton muscles, hearts and livers in KD-affected and non-affected areas are shown in table VI. It indicated that there are no significant differences between patients suffering from KD and other diseases in the endemic area. However, averages of selenium levels of hearts, livers and kidneys of patients in selenium-adequate areas are around 10-fold those from KD-affected areas. These results are in agreement with what was obtained previously from blood selenium concentration and GSHPx activity estimations.

Selenium Loading Test. The poor selenium status was also demonstrated by the urinary excretion of selenium after a loading dose of sodium selenite. In comparing the children in Beijing with those in the KD-affected area, it showed that 17% of the selenium dose was excreted in urine in 12 h in the former group, while only 4.3 and 1.7%, respectively, was excreted in children with and without previous sodium selenite supplementation in the latter group. The average hair selenium levels were accordingly 0.56, 0.23 and 0.061 ppm [Yin et al., 1979].

3. Prophylactic Effect of Selenium on Keshan Disease

Various prophylactic measures, each possibly implicating an etiological hypothesis of this disease, had been tested but none achieved convincing success until 1974 when a well-controlled large-scale trial with selenium was completed in Sichuan Province, southern China [Chen et al., 1980]. All children below 10 years old in the selected townships were under observation. They were divided into two groups, treated with a weekly oral dose of 0.5–1.0 mg of sodium selenite or with placebo for six successive months covering the prevalent season of this disease.

In 1974, 54 cases of various types of KD occurred among 3,985 subjects in the placebo group, i.e. 13.5‰. While in the selenium-treated group, there were only 10 cases observed among 4,510 children, i.e. 2.2‰. A similar study was conducted the next year. Again, 52 cases were seen among 5,445 placebo children and 7 cases among 6,767 selenium-treated subjects. The incidence rate was 9.5 and 1.0‰, respectively. The difference was highly significant ($p < 0.01$) between these two groups. This achievement confirmed the effectiveness of selenium for the prevention of KD and provided reliable evidence for linking selenium with a specific human disease.

This experience was widely adopted subsequently in China, and similar studies were conducted in many KD endemic areas. In Sichuan Province, there were more than 300,000 children who accepted selenium treatment in 1978–1982 (45,000–74,700 subjects per year and one person could be counted as 5 subjects in these 5 years). Among them, 88 cases of KD were observed. The average incidence rate was 0.27‰, with a variation of 0.15–0.47 from year to year. During the same period, however, the KD incidence rate observed in control subjects, children who had not been treated with selenium and who inhabited communities neighboring the selenium-treated area, was 1.55 on average, varying between 1.07 and 2.00‰, which is significantly higher than that of the selenium-treated populations [Xichang County Antiepidemic Station, unpublished].

A 10-year retrospective study which involved totally 100,000 subjects concluded that selenium supplementation reduced the KD morbidity from 10.71‰ in 1974 to 0.11‰ in 1983 [Wen et al., 1985] in Xichang Prefecture, a selenium-deficient zone in Sichuan Province. Meanwhile, evidence acquired from these studies indicated strongly a better heart function and prognosis of patients who have accepted supplementary selenium before the attack [Chen et al., 1980]. Reports from Heilongjiang, Shandong, Yunnan and Shaanxi provinces all lead to positive findings, and agreed well with the results obtained in Sichuan Province.

4. Therapeutic Effect of Selenium on KD

To test the therapeutic effect of selenium on this disease, 20 cases of acute or subacute type of KD were studied in 1976 [Taoping, 1978]. In addition to the curative measures conventionally employed for cardiomyopathy treatment, a daily oral supplementation of 0.5–1.0 mg of sodium selenite was given to each patient for the first 3 days since admission. After that, another three dosages were given, once every 3 days. Then the patients were switched to the preventive program, one dosage a week, for nine consecutive months. Findings of this study suggested that selenium supplementation brought about little benefit to the patients. In comparing with that of matchable control cases, the clinical signs, electrocardiography and X-ray examinations showed no more improvements in selenium-treated subjects, although their selenium concentration and GSHPx activity level of blood elevated obviously. The possible interpretation of this result could be that the elevated selenium concentration and GSHPx activity of blood are not powerful enough to correct the heart dysfunction resulting from severe myocardial damage. Clinical diagnosis of KD relies mainly on signs of cardiogenic shock and/or congestive heart failure. Whenever the symptoms and signs were severe enough to make a diagnosis, there is usually extensive myocardial damage existing in this organ. Myocardial necrosis is an irreversible change and can be repaired only by fibrous replacement associated with a loss to a certain extent of the heart function. It seems not reasonable to expect a complete recovery of the damaged heart by means of selenium treatment. This is not to say, however, that selenium is no use for KD therapy. Very few therapeutic studies have been launched to test the effectiveness of selenium, and there are too little data available to make a conclusion at present. Considering the insufficient liver function of the patients, a small dosage of selenium was used in the study mentioned above. Therefore, it might be underexposed. Furthermore, whether vitamin E should be given as an accompaniment with selenium is not clear.

A better prognosis of patients who have received supplementary selenium was revealed in the 1974–1975 study [Chen et al., 1980]. There were 17 cases of KD in the selenium-treated group. Of them, only one death occurred in the following 3 years. Of the 106 patients in the placebo group, however, 53 died during the same period. The heart function was significantly better in the selenium-treated survivors than the one in placebo cases. This result might implicate a later stage effect of selenium in KD therapy.

An excessive amount of selenium has been accidentally given to KD patients in a village clinic in the late 1960s. Remarkable improvements of heart function were observed in some of these cases with the appearance of hepatic toxication [Wang, 1972]. Therefore, whether or not a proper dose of selenium should play a positive role in KD therapy remains to be clarified.

5. Morphological Changes of KD

During the year of heavy prevalence of KD, more than 3,000 autopsy cases had been studied by 1960s [Lu, 1961]. The heart was found to be the main target organ. The principal pathological change is multifocal necrosis of the myocardium. The heart usually appears more or less ball-shaped with the dilatation of all chambers. Necrotic foci of various size and various stage of fibrosis are disseminated throughout the muscular walls. No substantial changes were seen in coronary arteries (fig. 3). Ultrastructural studies revealed two types of myocardial necrosis which agree well respectively with the contraction band necrosis and myocytolysis described by light microscopy. In general, the membranous structures, such as mitochondria or sarcolemma, are affected earlier than other organelles. For detail one may refer to Ge's review [Ge et al., 1983]. Pathologists believe that the morphological characteristics of KD heart should be referred to as ischemic necrosis. It is unlikely to be a result of coronary insufficiency but more likely to be based on metabolic disorders of the myocardium [Li et al., 1985b]. The fully developed heart lesions of KD are distinguishable from other congestive cardiomyopathy. Myocardial changes observed both in southern and northern China are of substantial identity but with recognizable differences. It might suggest a basic or principal pathogenic factor which is existing commonly all over the endemic zones and other coordinate or conditional factors which may vary from one endemic zone to

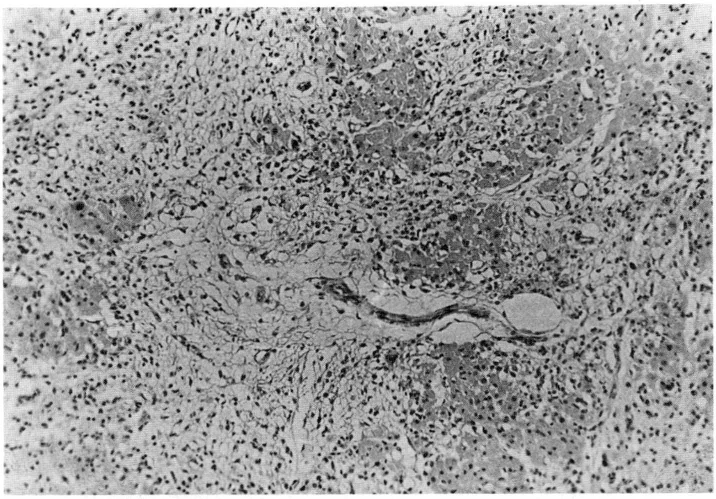

Fig. 3. Myocardionecrosis of Keshan disease. HE. × 120.

another. Long-term parenteral nutrition resulted in fatal myocardiopathy [Johnson et al., 1981; Fleming et al., 1982]. The selenium deficiency of these victims was demonstrated by extremely low selenium concentrations in blood and tissues. Postmortem examination of heart revealed widespread myocardial loss and replacement fibrosis resembling that of KD.

After selenium was proved to be related to KD, vigorous actions were taken to induce KD-like heart lesions in experimental animals, using the selenium-deficient diet alone or combined with other suspective risk factors. It seems right that selenium deficiency alone may not be strong enough to cause cardiac alterations unless a coexistant factor occurs. Looking into the experiments with positive pathological findings, a cofactor was always detectable in addition to selenium deficiency though it might have been overlooked sometimes. A close link of selenium deficiency with myocardial necrosis were regularly observed in suckling mice which were challenged with Coxsackie virus B4 [Bai et al., 1982]. It suggests strongly that viral infection may be a cofactor existing in some endemic areas.

A review of the animal experiments in this field indicated that the pathological features vary not only from species to species but also from study to study with the same animal species. Therefore, one should not expect an animal model bearing exactly the same morphological changes as that of KD human beings. One cannot either justify a KD animal model relying mainly on some morphological similarities when the real cause(s) of the disease is still obscure.

White muscle disease occurs spontaneously and experimentally in livestock due to selenium and vitamin E deficiency [Piglet Illness Research Group, 1976]. Lambs, calves and pigs are affected frequently in farms located within the KD-endemic zones in China. Sodium selenite administration protected the animals successfully from this disease [Veterinary Station of Jiusan Farm, 1976]. Both the similarities and the differences of morphological characteristics between white muscle disease and KD have been described in the literature.

Based on the morphological character of farm calves suffering from selenium-vitamin E deficiency cardiomyopathy, Li et al. [1985b] suggested that these affected animals may serve as a natural animal model for KD. In the same paper, however, the author admitted that selenium-vitamin E deficiency may induce changes of different characteristics in the same organ of the same animal species. Again, to justify an animal model by some non-specific morphological changes for a human disease so complicated as KD will hardly be accepted with enough confidence.

6. Clinical Manifestations of KD

The outstanding clinical feature of KD is cardiogenic shock and/or congestive heart failure which may vary from mild to severe. The next in frequency is arrhythmia and conduction block. Thrombosis and embolism may present in a small portion of cases. Four types could be categorized clinically.

Acute KD. It may occur abruptly in outwardly healthy people without preexistence of any cardiac disorders, or result from a rapid exacerbation of a chronic or insidious case. Dizziness, nausea, precardiac and/or substernal discomfort are the most common symptoms; vomiting, feeling chill, chest pain and dyspnea may also be complained of by some patients. Physical examination may reveal, in the early stage, signs mainly associated with cardiogenic shock, such as pallor, cold extremities and low blood pressure (often below 80/60 mm Hg). Later, signs of congestive heart failure become evident, when shock is controlled. ECG shows multiple variations and rapid changes of the waves, intervals and S-T segment. Roentgenograph examination indicates slight to moderate heart enlargement with diminished pulpation.

Chronic KD. It may occur spontaneously or be transformed from acute, subacute or insidious cases. The patients always complain of palpitation with consciousness of the heart beat at rest, short of breath, edema and oliguria. All signs associated with congestive heart failure can be detected through a physical and instrumental examination. A flask-shaped heart enlargement with weak pulpation is characteristic by X-ray examination.

Subacute KD. This type is most commonly seen especially in children. Its clinical feature is a combination of those of acute and chronic types. The onset should be milder than that of acute cases but more rapid than chronic cases. Facial edema, gallop rhythm and moderate heart dilatation are the typical signs of subacute patients.

Insidious KD. Patients usually show mild heart enlargement with normal heart function. They may complain of dizziness and palpitation after heavy physical labor. This feature often remains unchanged with no medication for years, and discovered by a routine checkup. Right bundle branch block, mild A-V block and ventricular extrasystoles may also be present.

The curative measures for chronic KD are of little difference from the generally accepted conventions for congestive heart failure. For the cardiogenic shock, intravenous administration of huge amounts of ascorbic acid has been used successfully since the 1960s, and dramatic effectiveness has been achieved in many local clinics. A repeated rush dose of 5–10 g ascorbic acid was employed before the subsidence of the syndrome. The total amount should not exceed 30 g daily by 3 days. Electrocardiography showed evidently that intravenous injection of large doses of vitamin C could eliminate the abnormal waves resulting from metabolic damage of the subendocardial myocardium. The blood concentrations of lactic acid and pyruvic acid were elevated sharply during the attack. Ascorbic acid administration may accelerate their catabolism and reduce their blood contents [Li et al., 1984b]. This will improve the biochemical environment and contribute to the functional recovery of the myocardium before it is transformed to the irreversible phase, necrosis. Treatment of a few cases of cardiomyopathy other than KD with ascorbic acid detected less obvious interests [Fang, 1975, unpublished]. Glucose transfusion has been used conventionally for releasing the shock syndrome. It seems effective but less remarkable than ascorbic acid administration.

7. Significance of Selenium Deficiency on the Etiology of KD

Results from epidemiological studies and selenium supplementation trials would reveal a causative association of KD and selenium deficiency, since the KD area is consistently in poor selenium status. However, the findings that poor selenium-status need not produce KD, and the fact that fluctuation in hair selenium concentration was not observed during the period of peak incidence, would suggest the multiple causality of KD. In view of the fact that the distribution of KD in China covered such a wide area as to include 15 provinces and autonomous regions and that all focally affected areas were consistently in extremely poor selenium status, as judged by selenium concentration in tissues, urine, blood and GSHPx activities in blood in comparison with the nearby non-affected areas, and the effective dose of selenium supplemented is a minute amount as such, i.e. 60 μg/day, it is obvious that the role of selenium added would be a normal physiological recovery rather than a pharmacological action, and selenium deficiency would be reasonably a fundamental factor necessary for the pathogenesis of heart muscle.

8. Other Hypotheses

Other hypotheses have been suggested on KD etiology. Here is a brief summary.

(1) Infection

KD is thought to be a kind of naturally occurring infection transmitted by unknown mediators or to be an infectious disease of viral origin. The latter suggestion is still considered to be the most probable factor, although no specific virus is yet known. Several strains of enterovirus, such as ECHO and Coxsackie, have been isolated from the blood and organs of KD patients [Su et al., 1979], but none of them could be repeatedly isolated from the patients or proved responsible for the pathologic change of heart muscle in KD patients.

(2) Chronic Intoxication

It has been suggested that KD may occur through chronic poisoning by environmental toxicants. Contamination of drinking water or foods or both by certain soil humus, natural plant toxins, soil inorganic chemicals, such as barium, nitrite, etc., or by mycotoxins in moldy grains has been suggested. No definite evidence of these toxicants has yet been found to be responsible, although extensive analyses have been conducted.

(3) Inadequacy of Nutrients Intake

Though protein and calorie intakes in the affected areas are generally comparable to those in the non-affected areas, protein or specific amino acid deficiency has been suggested as causative factors. Most foods eaten by residents in the affected areas are grown locally on land which is eroded. Furthermore, the dietary pattern in rural, affected areas is rather monotonous. Either dietary deficiency of molybdenum, magnesium or thiamine or imbalanced intake of electrolytes as well as certain trace elements has been suggested as causes of KD.

Unfortunately, none of these hypotheses has explained successfully the etiology of KD and controversial reports are still appearing. The demonstration that KD is closely related to selenium deficiency and selenium supplementation can reduce the incidence of KD thus afforded important information on the etiology of this disease. Because of the apparent multi-causes of KD in nature, forthcoming research carried out on the basis of combining selenium deficiency with other suspected factor(s), such as physiological stress, may be expected to secure more valuable information for the evaluation of the etiology of KD.

9. Other Possible Factors Involved in the Selenium Deficiency Hypothesis

(1) Virus

Experiments have been conducted to determine the combined effect of dietary selenium and viral infection on the myocardium of mice [Ge et al., 1984]. Weanling mice were fed with natural and semisynthetic low selenium diets. Gradient amounts of sodium selenite were given to animals. After 6 weeks on their experimental diets, the animals were bred within treatment groups. At 7 days of age, the offsprings were injected intraperitoneally with the Coxsackie B4 virus isolated from the blood of a patient with subacute KD. The animals were sacrificed one week later. The incidence of heart lesions, as revealed by pathological examination, was much greater among the selenium-deficient mice (31–40%) than that in the selenium-supplemented group (10–23%) ($x = 54.4$, $p < 0.00001$) (table VII). An inverse correlation between blood selenium concentrations and incidence of myocardial lesions was observed ($r = -1.789$). It was concluded that selenium treatment protected selenium-deficient mice against the induction of heart lesions by Coxsackie B4 virus.

(2) Vitamin E

Either adequate or marginal levels of plasma vitamin E has been reported in general population of endemic areas [Yang et al., 1984c; Li et al., 1983; Hou and Zhu, 1982]. No significant differences of plasma vitamin E levels either between patients and normals or between residents in affected and non-affected areas were found. Since low vitamin E intake

Table VII. The influence of supplemental selenium (Se) on myocardial necrosis induced by viral infection in mice. From Ge et al. [1984]

Group	Dietary Se, ppm	Blood adult	SE (ppm) suckling	Total n	Heart n	Lesions %
Experiment A						
Sichuan	0.010	0.053	0.032	56	20	36
Sichuan + Se	0.148	0.183	0.138	30	4	13
Stock	0.518	0.402	0.306	52	8	15
Experiment B						
Sichuan	0.017	0.054	0.033	44	17	38
Sichuan + Se	0.147	0.150	0.070	58	8	14
Semisynth.	0.015	0.049	0.024	74	23	31
Semisynth. + Se	0.145	0.111	0.088	70	16	23
Stock	0.334	0.440	0.321	72	8	11
Experiment C						
Sichuan	0.027	0.075	0.059	182	73	40
Sichuan + Se	0.062	0.210	0.146	97	19	20
Sichuan + Se	0.462	0.582	0.316	128	13	10
Semisynth.	0.027	0.130	0.070	68	22	32
Semisynth. + Se	0.062	0.255	0.146	71	14	20
Semisynth. + Se	0.462	0.599	0.328	74	9	12

A natural low Se diet consisting mainly of cereals produced in the Keshan disease endemic area of Sichuan province. Se provided as sodium selenite by stomach tube. A stock laboratory diet used routinely in our Institute. Se added as sodium selenite to drinking water. A semisynthetic diet based mainly on starch and low Se yeast. Estimate based on calculation of total Se intake from Se content of ingredients. Indicating a statistical difference between Se-treated and untreated groups by χ^2 analysis.

may increase selenium requirement and influence the pathological development of heart lesions [Reinhold, 1975], particularly in the case of selenium insufficiency, it seems inappropriate to exclude the possibility that marginal vitamin E might under certain conditions act as a cofactor with selenium deficiency to accelerate the development of the pathological changes of heart muscle.

(3) Protein and Methionine

It was found that when a selenium-deficient diet is fed to animals, any increase in dietary protein levels will consequently stimulate animal growth and further reduce the selenium concentration of animal tissues

[Zhou et al., 1983; Sun et al., 1983]. It showed a significant negative correlation between animal body weight and tissue selenium concentration, but a positive correlation was obtained if the total amount of selenium contained in the whole organ was used instead of tissue concentration. Obviously there is no favorable effect on the utilization of dietary selenium if protein alone, rather than protein foods is supplemented to the low selenium, endemic diet. In other words, it is unlikely that protein-calorie deficiency might play any role in the pathogenesis of KD. However, as methionine was added to the endemic diet at a level of 0.13%, the GSHPx activity and selenium content in the liver increased significantly. It suggested that the endemic diet may be somewhat deficient in methionine. Mertz [1982] suggested the possibility that the second conditional factor which allows the KD to appear might consist of the slfur-containing amino acids. It was also reported that an elevation of plasma and erythrocyte GSHPx activities was observed as methionine was supplemented to the endemic diet in a low selenium area in China [Luo et al., 1984].

10. Possible Mechanism of Pathogenesis

(1) Impairment of Energy Metabolism

Early studies observed that the characteristic morphological changes in the myocardium of KD patients somewhat resemble the changes caused by ischemia and hypoxia. It was suggested that the pathologic change of heart muscle might be due to impairment of energy metabolism from lack of oxygen [Wang, 1962]. Morphologic observation by electron microscope revealed that the early and conspicuous change occurred in the mitochondria of myocytes [Ge et al., 1982]. This observation again suggests the failure of energy metabolism in heart tissue. Later work found that activities of some respiratory enzymes such as succinic dehydrogenase and cytochrome *c* oxidase in autopsy heart tissues of patients were severely reduced [Yang, 1982].

(2) Damage of Biomembrane

It has been shown that selenium is a component of GSHPx [Rotrack et al., 1973]. In the case of KD patients with too low activities of this enzyme in their heart muscle, biomembrane in cells may be expected to be damaged due to uncontrollable peroxidation in membrane phospholipids. Histochemical studies on the one hand revealed that the activity of acid phosphatase was obviously increased in muscle fibers surrounding the necrotic foci in the myocardium of the KD patients [Ge et al., 1982]. The authors suggested that the membrane function of organelle would be altered. On the other hand, it was found that the erythrocyte phospholipid content of KD patients was reduced [Li et al., 1984a]. In in vitro studies, it was found that the hemoglobin in the intact red cells of inhabitants in the KD area is more susceptible to the peroxidant effect of the autoxidation of hydroxyfumaric acid [Zhu et al., unpublished]. The above observations demonstrate that the blood cell membrane of patients suffering from KD is damaged. Working with mitochondria it was found that the activities of H^+-ATPase and related respiratory enzymes linked to the membrane of mitochondria isolated from heart tissue of KD patients were reduced. The conclusion was thus drawn by the authors that both the structure and function of mitochondria membrane are altered in KD patients [Yang et al., 1986, unpublished]. Since GSHPx may not be able to protect membrane through the reduction of esterified fatty acid hydroperoxides which might occur in membrane [McCay et al., 1976], it was suggested that vitamin E functions as a specific lipid-soluble antioxidant in the membrane and selenium functions as a component of cytosolic GSHPx that reduces peroxides [Noguchi et al., 1973]. Thus, GSHPx and vitamin E act in a complementary fashion to block free radical formation which subsequently attack the membrane phospholipids. Since no conclusive evidence has been obtained that the KD patient is consistent with vitamin E deficiency, the mechanism of the biomembrane damage observed in KD patient remains to be elucidated. Recent research, however, found that the effect of selenium to stabilize and protect biomembrane seems not necessarily through GSHPx, but through an unknown form of selenium binding to the membrane [Wo and Yang, 1986]. Whether or not this discovery could help explain the mechanism of heart muscle biomembrane damage of KD patients awaits further investigation.

B. Kashin-Beck Disease

1. History

KBD is an endemic osteoarthropathy, showing atrophy, degeneration and necrosis of cartilage tissue. Description of this disease was first made by I.M. Yurensky in 1849 and later by N.I. Kashin (1859–1860) and E.V. Beck (1899–1906) in detail in the Urov river area of Siberia in the USSR. Results from later research supported the hypothesis that KBD is a biogeochemical disease in its etiology, and this disease is now under control in the USSR.

About 300 years ago, a likely description of this disease appeared in the county records of Anze, Shanshi Province, in northwest China in 1644, a heavy prevalence county of KBD today, but a detailed description was not made until 1934 by Zhang Fengshu in Jilin Province, northeast China. KBD is now prevailing in 14 provinces, municipalities and autonomous regions of mainland China. A survey made in 1950 in north China indicated that the incidence rate generally fell into the range of 30–40% and mounted to 70% in certain localities. A more recent survey showed a very high incidence of 96% in a heavy prevalence village in the southwest of China. The distribution of KD and KBD in most cases is overlapped in the north of China. Like KD, KBD is distributed within a wide zone running from northeast to southwest of China which was demonstrated as a selenium-deficient zone [Tan et al., 1982]. Following the documentation of the close relationship between selenium deficiency and KD, reports appeared immediately to confirm the same relationship between selenium and KBD [Li, 1981; Hou and Zhu, 1982]. Oral selenite-vitamin E therapy was early reported effective for the cure of KBD [Li, 1979]. Later work indicated that sodium selenite alone was also effective [Liang, 1984a, b]. However, a less distinct effect has also appeared [Deng, 1985, unpublished]. The effect of KBD prevention by selenium needs to be further confirmed.

2. Epidemiology of KBD and Selenium

Epidemiologically, the main feature of the regional characteristics of KBD is similar to KD. In heavily affected areas, new cases of KBD can be detected at 2–3 years of age by X-ray examination whereas in mildly affected areas the initial age of onset was usually postponed beyond 10 years. Generally the vulnerable population are children within the age of 5–13 years old. Very few cases are seen in adolescents and adults. In contrast to KD, KBD appeared not only in families of peasants but also in families of herdsman and children from government employees although the prevalence is usually milder in the latter case. It seems that the dietary 'quality' required to prevent KBD is somewhat higher than that of KD. The distribution of KBD is less stable and it may appear occasionally in certain localities within the low selenium area which has no cases reported before.

(1) Regional Distribution and Selenium Status

Since the close relationship between selenium-deficiency and KD has already been documented, and the occurrence of KD and KBD frequently accompanied each other in the same area, it is crucially important to

Table VIII. Hair selenium concentration (ppm) of residents in Kashin-Beck disease-affected and nearby non-affected areas in Yulin District

Site	Affection	Number of subjects	Concentration
Erlintu	+	19	0.081 ± 0.011
Niujialiang	+	25	0.077 ± 0.008
Mizi	–	25	0.195 ± 0.011

Results are expressed in mean ± SE.

Table IX. Hair and blood selenium concentrations (ppm) of Kashin-Beck disease-affected and nearby non-affected populations in Aba Automous Region

Sites	Affection	Hair	Blood
Agricultural area			
Yanyun, peasants	+	0.081 ± 0.002 (42)	0.016 ± 0.002 (10)
Gontu, peasants	–	0.144 ± 0.014 (20)	–
Yanyun, staff	–	0.251 ± 0.054 (22)	–
Stock-raising area			
Caoyuan, herdsman	+	0.165 ± 0.007 (41)	0.021 ± 0.002 (16)
Caoyuan, staffs	–	0.357 ± 0.019 (20)	–

Numbers in parentheses are the number of subjects examined. Results are expressed in mean ± SEM.

examine the selenium status in areas where KBD occurred alone. Results obtained are shown in tables VIII and IX. It indicated that the selenium status of the general population in sole KBD areas, such as Yulin District, northwest China, and Aba Autonomous Region, Southwest China, were invariably poor [Yang, 1984a]. It is shown in table IX that the hair selenium level of the population in Yangyun, the affected site, was 0.081 ppm, while in the non-affected site, Gontu, it reached 0.144 ppm, but the hair selenium level of the population in Caoyuan even reached 0.165 ppm and KBD was still prevalent. Probably the intensity of stress is different between the two sites. It is worth noting here that KBD-affected areas are always classified according to the activities of the local etiological factor(s) through bone X-ray examination and blood enzyme activity estimation by the Chinese workers.

(2) Selenium Levels of Foods and Human Tissues [Yang, 1984a]

Staple Cereals. The distribution of selenium in cereals in KD, KD-KBD, KBD and non-affected areas is shown in table X. Each figure shown is the average of samples taken from different sites over China. It showed that the averages obtained from areas of KD, KD-KBD and KBD are similar and are markedly lower than those of the non-affected area. Since rice produced in the endemic area in northeast China usually contained a higher concentration of selenium than wheat and corn produced in the same area, as reported in section IV.A.2 (1), rice data from northeast China were excluded from the table.

Blood and Hair. The distribution of selenium in blood and hair is shown in tables XI and XII, respectively. Although the averaged blood selenium concentration was around 0.020 ppm in endemic areas (table XI), a considerable amount of blood samples in the endemic area were extremely low. For instance, 325 inhabitants in KD and KD-KBD areas averaged 0.018 ppm in which 22% were below 0.01 ppm and averaged 0.007 ppm. Averages of hair selenium concentrations in each site obtained in the KD area were below 0.123 ppm (table XII), while near the KD-affected areas they ranged from 0.123 to 0.165 ppm and that apart from

Table X. Selenium concentration (ppm) of staple cereals produced in KD, KD-KBD, KBD and non-affected areas. From Yang [1984a]

Area	Corn	Wheat	Rice	Soybean
KD	0.006 ± 0.001 (58) (0.001 – 0.018)	0.008 ± 0.000 (59) (0.003 – 0.019)	0.008 ± 0.000 (59) (0.002 – 0.020)	0.014 ± 0.002 (51) (0.002 – 0.043)
KD-KBD	0.004 ± 0.000 (143) (0.001 – 0.012)	0.006 ± 0.000 (72) (0.001 – 0.018)	–	0.008 ± 0.000 (98) (0.001 – 0.031)
KBD	0.005 ± 0.000 (24) (0.002 – 0.008)	0.006 ± 0.001 (26) (0.002 – 0.014)	–	–
Nonaffected	0.029 ± 0.004 (79) (0.005 – 0.286)	0.038 ± 0.004 (110) (0.007 – 0.417)	0.043 ± 0.005 (44) (0.009 – 0.195)	0.063 ± 0.011 (40)

Results are expressed in mean ± SE; numbers in upper parentheses indicate the number of samples analyzed; numbers in lower parentheses indicate the range. KD-KBD = Keshan disease + Kashin-Beck disease.

endemic areas where they were higher than 0.2 ppm. As data from KBD areas were included, the upper limit of the affected area was extended to 0.165 ppm (table XII).

3. Clinical Manifestations of KBD

The disease occurred mainly in children of 5–13 years old and very infrequently in adolescents and adults. The incidence rate was not different between the sexes. As described by Yin [1985] and Huang et al. [1984], fatigue and joint stiffness were the most common symptoms at the beginning. Limitation of flexion of fingers toward the palm and elbow joint movement appeared gradually. A stabbing pain was noted often in weight-bearing joints, and this was aggravated by being exposed to cold and/or

Table XI. Selenium concentrations (ppm) of whole blood of residents in KD, KD–KBD, KBD and non-affected areas. From Yang [1984a]

Areas	Number of subjects	Blood selenium concentrations	Range
KD	210	0.019±0.001	0.002–0.062
KD-KBD	115	0.015±0.001	0.003–0.036
KBD	25	0.019±0.002	0.006–0.039
Non-affected	134	0.093±0.007	0.013–0.351

Results are expressed in mean ± SEM.
KD-KBD = Keshan disease + Kashin-Beck disease.

Table XII. Average hair selenium concentrations (ppm) among populations in KD, KD-KBD, KBD and non-affected areas. From Yang [1984a]

Areas	Number of sites	Number of subjects	Average	Range
KD	39	697	0.077±0.004	0.027–0.123
KD-KBD	36	781	0.068±0.004	0.034–0.113
KBD	5	167	0.094±0.03	0.035–0.165
Non-affected	111	1,843	0.361±0.015	0.124–0.855

Results are expressed in mean ± SEM.
Two sites (0.094 and 0.089) were excluded.
KD-KBD = Keshan disease + Kashin-Beck disease.

exercise. Enlargement and deformity of joints were characteristics in advanced cases, and muscular atrophy could be seen during this course. In severe cases, underdevelopment of stature may occur, and this should be more obvious if children were affected during their early life.

In the summary report of the IPCS/WPRO Meeting on Kashin-Beck Disease (Beijing, 1985), it was clinically classified as follows: *Early stage:* Flexion terminal part of fingers or bow-like fingers, pain of knee and ankle joint and doubtful joint enlargement and stiffness in the morning. *First degree:* Enlargement and crepitus of small joints, limitation of elbow, wrist and ankle joints. *Second degree:* Short fingers, enlargement and dysfunction of medium joints. *Third degree:* Enlargement and dysfunction of large joints, underdevelopment of stature.

4. Pathological and Roentgenographic Changes of KBD

The basic pathological changes in KBD are multiple degeneration and necrosis of hyaline cartilage tissue and repairs on it. The epiphyseal cartilage, articular cartilage and epiphyseal growth plate are the mainly affected sites. The degenerative changes appear first at the deep layer of the articular cartilage and involve the hypertrophic chondrocytes of the epiphyseal plate. In severe cases, lesions may expand to the surface zone of the cartilage as well [Mo and Lue, 1985].

In the necrotic area, the chondrocytes dies, one batch after another. The chondrocyte nuclei diffuse first, then the cytoplasm became faint or eosinophylic in stain and the cells became ghosts. Later, the degenerated cells disappeared completely and the affected area became homogenous. The necrotic foci of various size were observed both in the articular and epiphyseal cartilage. The extremity bones are most often affected and, usually, a number of joints are involved at the same time [Mo and Lue, 1985].

Secondary to the chondronecrosis, proliferation of the chondrocytes was soon observed in the adjacent circumference. Secretory vesicles and rough-surface endoplasmic reticulum were evident in these newly developed cells. Dysosteogenesis, absorption of necrotic material, organization and ossification take place gradually and, eventually, the necrotic area was repaired through reconstruction of bone. Early ossification of the epiphyseal plate terminated the growth of bones and led to short fingers, short limbs and even dwarfism [Mo and Lue, 1985].

Ultrastructural studies revealed two types of chondrocyte changes, condensation and vacuolization of the cytoplasm. Calcium granules depos-

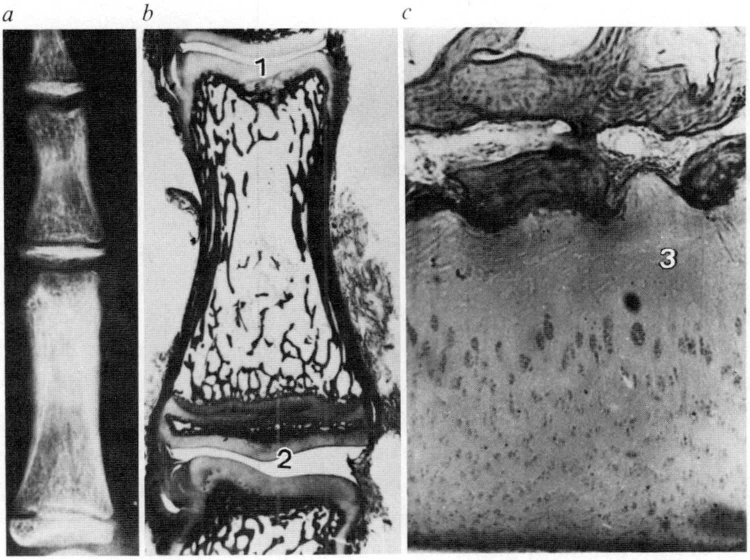

Fig. 4. Cartilage lesion of Kashin-Beck disease. *a* Interruption of the bone articular surface. *b* Necrosis of articular cartilage (1) and of that of the epiphysis (2). *c* Band-like cartilage necrosis (3) of the epiphysis. With the courtesy of Dr. Wang Yun-zhao, Beijing Jishuitan Hospital, China

ited in the cytoplasmic membrane. An increased number of intracytoplasmic filaments and myelin figures, a decreased number of membranous organelles and mitochondria distorsion were evident, indicating the degeneration of chondrocytes [Mo and Lue, 1985].

Chondronecrosis is the principal pathological change of this disease. X-Ray examination, however, cannot reveal it until the irregular calcification and ossification become evident as the consequence of absorption and repairing of the necrotic tissue (fig. 4). The basic X-ray signs of this disease were identified in connection with pathological changes [Wang et al., 1984a, b] and were summarized again in the summary report of the IPCS/WPRO Meeting as follows: (1) the calcification line becoming blurred, thin, broken, and disappearing as characteristic of chondronecrosis; (2) marginal defects of the metaphyseal plate, end of distal bone, and carpal and metacarpal bones following chondronecrosis; (3) deformity of epiphysis, synostosis of the epiphyseal plate resulting from necrosis of the whole layer of the epiphyseal plate; (4) enlargement of joints and stubby fingers as the effects of secondary osteoarthrosis.

5. The Prophylactic and Curative Effect of Selenium on KBD

The curative effect of selenium on KBD was first reported in 1979 [Li, 1979]. Patients were treated with oral administration of 0.5–2.0 mg sodium selenite weekly associated with vitamin E injection for a period of 3–6 months. Out of 224 cases, 187 (83%) showed improvement of the metaphyseal lesion in X-ray films. Niu et al. [1984] reported the preventive and curative effect of sodium selenite administration in Heilongjiang Province. The metaphyseal changes improved in 54 of 56 cases (96%), and the metaphyseal recovering rate was 86% (48 out of 56 cases). It was stated in the report that there were no new cases observed in selenium-treated children, but the sample size seems not big enough to make any conclusion. Studies in Shaanxi Province suggested strongly a therapeutic effect of a weekly oral supplementation of 1–2 mg sodium selenite [Liang, 1985]. The metaphyseal film showed improvement of various degrees in 267 of the 325 selenium-treated cases (82%). In the control group, however, 40% improvement and 19% deterioration were detected.

For prevention, selenium-fortified table salt was supplied to subjects from 1981 to 1985 in Shaanxi Province [Liang, 1985]. The incidence rate reduced from 65.9 to 46.8% in the selenium-treated group, and from 75.7 to 54.6% in the control group. No significant difference was detected between these two groups.

Yin et al. [1985a] sprayed the blossoming wheat crops with sodium selenite and elevated the grain selenium content by three times. The X-ray detection rate of abnormal metaphysis in children who had consumed the selenium-fortified wheat for 2 years reduced from 63 to 42%, and the abnormal metaphysis detection rate from 59 to 32%. In the control group, changes of both rates were not significant.

6. Possible Mechanisms of Pathogenesis

(1) Abnormal Metabolic Changes

Typical animal models of KBD (and KD) have not been found or produced although a tremendous amount of work was done. It has been revealed earlier that the incorporation of ^{35}S-sulfate into mucopolysaccharide excreted in the urine is reduced in KBD patients [Mo and Lue, 1985]. The reduction of chondroitin sulfate in the cartilage matrix of KBD patients was demonstrated due to the impairment of sulfation of chondroitin [Yang, 1984b]. An earlier report indicated that after the administration of ^{75}Se-selenate to suckling rats, radioactivity was detected in the long bones and cartilage. Histochemical study revealed that the radioactivity was associated with protein rather than with the polysaccharide component of the cartilage in rats. Later, working with rabbits, it was found that ^{35}S-sulfate was readily incorporated into the mucopolysaccharides of cartilage, whereas the entry of ^{75}Se from selenite or selenate into this fraction was almost non existant [Bopp et al., 1982]. It seems that the roles played by ^{35}S and ^{75}Se in cartilage metabolism are different. Whether or not ^{75}Se-deficiency will influence ^{35}S metabolism in cartilage tissue seems worthy of further investigation.

(2) Membrane Damage and Hypothesis

It was found that the morphology of erythrocytes of KBD patients is abnormal and both the phospholipid content and composition of the cell membrane are altered [Li, 1984]. Furthermore, the marked increase in Na^+, K^+, -ATPase activity and decrease in membrane fluidity would suggest an alteration in both structure and function of the membrane [Yang et al., 1984a, b].

The extremely poor selenium status of both KD and KBD patients have been well documented. In addition, the etiology of both diseases was suggested to be multifactorial. Beside selenium deficiency, imbalance of metals (Mn, P, Mo or Ba), mycotoxin, microbial toxin, virus, humic acid, etc. were suggested to be involved. The actual conditions for the development of KD or KBD or both in certain localities or subjects seems a matter of complexity. However, the possibility that membrane damage caused by selenium deficiency would weaken the resistence of the body to any environmental stress and thus provide a common basis for the pathogenesis of KD and KBD cannot be excluded. It has been proposed that selenium-deficient duck produced damaged fibroblast membranes, which resulted in decreased collagen synthesis [Brown et al., 1982]. In the KBD patient, it was found that the amount of hydroxylysine excreted in the urine is elevated and this was proposed as an indication of abnormal collagen metabolism [Yang, 1984]. A possible link between membrane damage and heart muscle necrosis has already been mentioned.

7. Prevention of KD and KBD in Practice

It was found that the selenium content of most staple cereals in endemic areas contained selenium less than 0.02 ppm (table X). The daily selenium intake of adults was usually around 10 μg or less in endemic areas and could only meet one fourth to one seventh of the lower limit of the selenium safe intake suggested by the US National Research Council

[FNB/NRC, 1980]. Perhaps more important is the fact that selenium intervention is effective for the prevention of KD. Although controversial results from selenium intervention were reported with KBD, a majority of the studies showed a positive response. Furthermore, patients suffering from KBD showed lowered blood GSHPx activities. All this suggests that it is reasonable to put selenium intervention into practice in KD and KBD endemic areas. Like KD, KBD was also proposed to be multifactorial in its etiology. Measures practised in China for the prevention of KBD were somewhat similar to KD in general, and had been reported to be effective somehow.

(1) Comprehensive Prevention

Stress was laid on personal and environmental hygiene including killing house rodents and purifying drinking water by filtration, precipitation or dropping bleaching powder into the well, or drinking deep well water. It is crucially important to encourage people to diversify their foods both in variety and source. For instance, the exchange of staple grains among different areas and the increase in the intake of soybeans and animal foods in order to get more selenium, vitamin E, methionine and other nutrients.

(2) Selenium Intervention

Three methods have been proposed for the application of selenite [Yang et al., 1982b]. They are: selenized table salt, foliar spraying selenite solution and selenite tablet. Since endemic areas are always eroded, it seems that care should be taken to avoid excessive supplementation of a single element. The goal of selenium supplementation in this country is only to elevate the poor selenium status of inhabitants to a level compared to the near non-affected areas, i.e. selenium concentration of hair >0.2 ppm of blood >0.06 ppm.

Selenite Tablet. The doses of sodium selenite used per week are 0.5 mg for small children within the ages of 3–5 years, 1.0 mg for children of 6–10 years and 2 mg for subjects older than 10 years of age. It is difficult for inhabitants to keep the oral supplementation regularly during long-term treatment unless a special delivery system to households is set up. In order to simplify the procedure, an alternative method of dosage system was used: tablets are given daily for the initial seven consecutive days. The daily dose was the same as used per week by the original method. Thereafter, the same dose is given only once per 2 weeks [Yang et al., 1982b].

Selenized Table Salt. The table salt was proposed to contain 10–15 ppm of sodium selenite [Yang et al., 1982b]. The advantages of using table salt as a carrier are due to its simplicity, cheapness and acceptance. Since the actual amounts of individual selenium intake are influenced by the dietary habit, the children, for instance, taking less dishes will consume less supplemented selenium. In a selenium intervention study using selenized table salt, it was proposed that sodium selenite concentration should be set at 15 ppm in order to meet the requirement of baby children [Qian et al., 1981]. This consideration is of importance. Mertz [1977] has pointed out earlier that the most important consideration in the selection of a carrier food must be its availability to the target population. Another problem seems the incapability of inorganic selenium salt to raise the body selenium store to a higher level. Even after long-term supplementation with 60–100 μg selenium daily, hair and blood selenium levels of subjects still remained around 0.2 and 0.06 ppm, respectively.

Foliar Spraying Selenite Solution. It has been recommended that 7.5 and 6.0 g sodium selenite applied per acre of wheat and corn respectively will raise the hair selenium concentration to an average of 0.3 ppm [Yang and Du, 1983]. The bioavailability of selenium in grains of sprayed crops is similar to that in natural products [Yang et al., 1982b]. Although selenium recovered from kernels of sprayed crops is rather low – less than 2% – it is still considered to be economical, however, if other parts of the crops are utilized as selenium-enriched feed for livestock. Foliar spraying seems to be the method of choice, but the ultimate decision still depends upon the feasibility of the method in different endemic areas.

V. Selenium and Cancer

A. The Relationship between Selenium Status and Cancer Risk in China

In recent years, there is a growing interest in the relationship between selenium and cancer, either in the field of nutrition study or cancer research. Among the dietary factors under current investigation, selenium seems to be one of the most promising nutrients, in terms of its relationship with cancer risk [Willet, 1984]. This is because the anticarcinogenic effect of selenium is consistently shown in both animal experiments and epidemiological studies by a number of authors [IPCS Environment Health Criteria-Selenium, 1986].

1. Ecological Studies

Until recently, most of the epidemiological evidences were based on ecological studies mainly carried out in western countries, although three prospective studies have been published very recently [IPCS Environmental Health Criteria-Selenium, 1987]. In these ecological studies, there is a consistent trend that crop selenium concentration (and selenium intake) as well as blood selenium content were reported to be inversely correlated with the mortality rate of a number of cancer sites. However, these studies have been criticized for having uncontrolled confounding factors, such as: (a) the insufficient correlation of the selenium status of local residents with the selenium concentration of locally produced crops [Allaway, 1972] because of the diversity of food sources; (b) the ill-defined migration habits of the local population; and (c) the large number of other dissimilarities in the areas investigated, in addition to selenium [Allaway, 1972]. In contrast, China offers a unique setting for the study of interrelationships between selenium status and cancer mortality for the following reasons. The majority of the Chinese population is located in rural areas where migration is very limited. They live on rather simple dietary patterns mainly comprised of locally produced foods. In addition, there is a wide variation of cancer mortality within China [Atlas of Cancer Mortality in the People's Republic of China, 1979], thus providing an ideal model for ecological studies.

Yu et al. [1985] reported that the selenium levels of whole blood from healthy adult subjects in 24 regions were inversely correlated with the age-adjusted total cancer mortality for both sexes ($r = -0.64$, $p < 0.01$ for males; $r = -0.60$, $p < 0.01$ for females). When the data on the selenium levels were compared with the age-adjusted mortality rates by organ sites, statistically significant negative correlations were found, particularly for the major cancer sites in China, i.e. esophageal cancer ($r = -0.66$, $p < 0.01$ for males; $r = -0.6$, $p < 0.01$ for females) and stomach cancer ($r = -0.64$, $p < 0.01$ for both sexes). When the blood selenium levels were classified into low, medium and high selenium areas, significantly lower total cancer death rates were observed in areas with high selenium levels and the mortality rates of esophageal, stomach and liver cancers were particularly increased in the low selenium areas (fig. 5). In the same paper, Yu et al. [1985] showed that in Qidong County of the Jiangsu Province, one of the high-risk areas for liver cancer in China, significant inverse correlations were found between the selenium content in maize and barley corn and the liver cancer mortality rates of the 43 communes (table XIII). Maize and

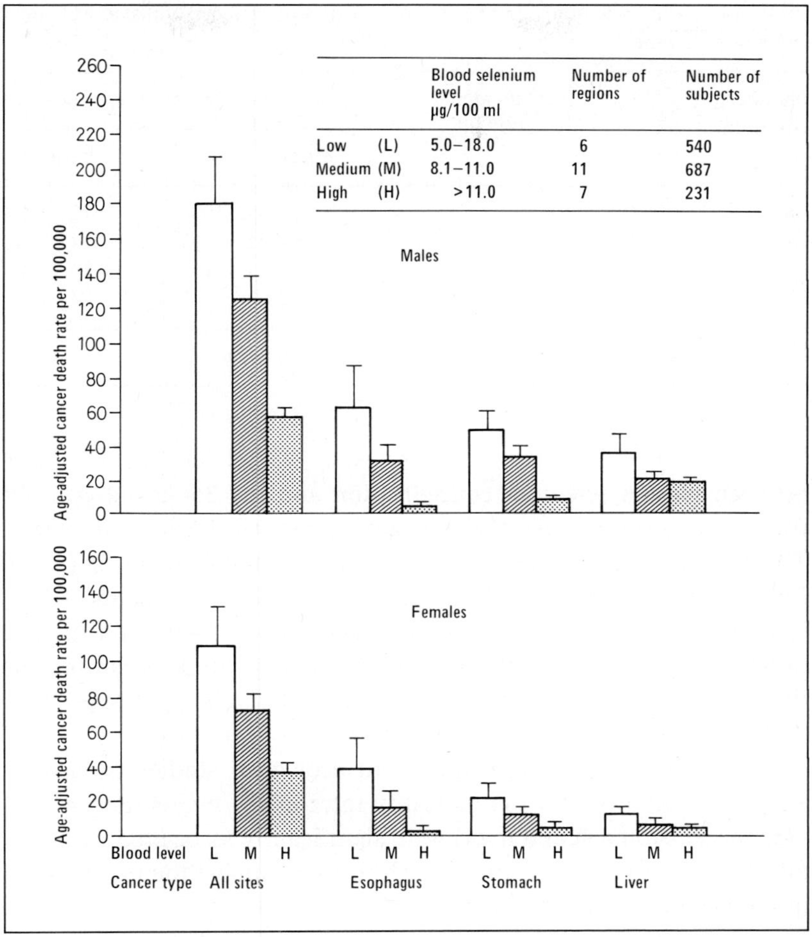

Fig. 5. Relationship between blood selenium level and age-adjusted cancer mortality per 100,000 population (mean ± SE). From Yu et al. [1985].

barley corn are the main staple foods in Qidong County and most of the dietary selenium in the local diet is derived from these two crops.

Chen et al. [1986] carried out an ecological study in China covering 65 counties with wide ranges in cancer mortalities for seven major cancer sites; and the relationships between a number of nutrients, including selenium, and cancer mortality were studied. Although the study has not been completed, preliminary results show that an inverse correlation was found

Table XIII. Grain selenium content in regions with different age-adjusted PLC incidence in Qudong County

Number of communes	PLC incidence per 100,000	Selenium content (ppm), mean ± SD	
		maize	barley corn
8	15–39	15.1 ± 4.4 (40)	42.3 ± 1.9 (39)
17	40–49	13.5 ± 2.2 (85)	24.7 ± 7.8 (84)
17	50–59	12.7 ± 1.6 (81)	23.5 ± 4.9 (80)
1	60.6	9.7 (5)	18.1 (5)

Numbers in parentheses are the number of samples analyzed.

between plasma selenium concentration and the 35- to 64-year-old age truncated mortality rates of esophageal and stomach cancer in both sexes. The correlation coefficients were −0.44 ($p < 0.001$) and −0.28 ($p < 0.02$) for esophageal cancer in males and females, respectively, and −0.28 ($p < 0.02$) and −0.39 ($p < 0.001$) for stomach cancer. Studies on the interactions between selenium and other nutrients (e.g. lipids, vitamins, dietary fibers, etc.) as well as several environmental carcinogens (e.g. *N*-nitroso compounds, aflatoxin, etc.) are under way.

The above results obtained from ecological studies carried out in China have shown that the mortality rates of the major cancer sites (i.e. esophageal, stomach and liver) were significantly associated with the selenium status of local residents, which further support the previous studies in the USA and other countries.

2. Intervention Studies

In order to confirm the role of selenium in some major cancer sites in China, several intervention studies are under way in certain high-risk areas. In all the studies, physiological or nutritional doses are being used rather than pharmacological doses. For example, a well-controlled intervention study on esophageal cancer provided 50 μg of sodium selenite daily to 1,700 subjects with precancerous lesions and to 15,000 general population subjects 40–69 years old; this was initiated in 1985 in Linxian, under the direction of Dr. J. Li of the Cancer Institute of the Chinese Academy of Medical Sciences and with the funding support from the American National Cancer Institute [Li et al., 1985a; Blot and Li, 1985].

Another intervention study on liver cancer was started in 1984 in more than 300,000 people in the above-mentioned Qidong County, directed by Dr. S. Yu of the Cancer Institute of the Chinese Academy of Medical Sciences and Dr. L. Clark of the Cornell University, USA. They are using table salt, fortified with 15 ppm sodium selenite. Since the 40- to 69-year-old age truncated mortality rates of esophageal cancer in Linxian and the male age-adjusted liver cancer mortality rates in Qidong County are as high as 470 and 76.26 per 100,000, respectively, it is expected that some preliminary results will be seen within 5 years.

B. Possible Mechanisms of the Anticarcinogenic Effects of Selenium Studied

A series of in vivo and in vitro studies have been carried out by Yu and her colleagues, in order to elucidate the possible mechanisms of the anticarcinogenic effects of selenium. Selenium was shown to have a selective effect on adenosine 3′:5′-cyclic phosphate (cAMP) metabolism in hepatoma cells and inhibit glutamate and pyruvate oxidation by using a mitochondria system prepared from hepatoma cells, as well as the swelling of mitochondria, but did not affect the phosphorylation of mitochondria isolated from normal liver cells [Yu and Wang, 1983; Yu et al., 1985]. The administration of sodium selenite remarkably depressed lipid peroxidation in Walker carcinoma 256-bearing rats as shown by the decrease of serum lipid peroxide concentration and a synergistic effect was observed when tocopherol and selenium were given simultaneously [Yu et al., 1984a].

VI. Possible Adaptation to Low Selenium Intake and Study of Selenium Requirements of Humans

A. Evidence as to the Possible Existence of Adaptation to Low Selenium Intake

Epidemiologic surveys before the 1950s indicated that the prevalence of KD was a very severe problem not only due to its high fatality, but also its high incidence. However, at present its epidemiologic characteristics has been markedly changed. On the one hand, the prevalence of this disease became far milder, i.e. the average morbidity of acute and subacute cases in north China decreased from 40 per hundred thousand in 1959 to 3.6 in 1981 and still progressively decreased to the present historical minimum record even though in some of the affected areas the poor selenium status may be similar. On the other hand, newly affected areas either expanded from the areas nearby or were recently discovered in the 1960s where only children got sick [Yuan, 1983]. Such a noticeable change in the

Table XIV. Mortality of Keshan disease (KD) in local and foreign immigrants, and its relation to the duration of residence in Beian County from 1953 to 1956

Duration of residence years	Foreign immigrants			Immigrants from near low selenium areas			Local residents		
	total[1] population	died of KD	mortality[2] (per 1,000)	total[1] population	died of KD	mortality[2] (per 1,000)	total[1] population	died of KD	mortality[2] (per 1,000)
<3	941	12	13	470	3	6	–	–	–
3–10	1,564	102	61	1,012	10	10	–	–	–
>10	2,158	34	16	4,364	24	8	–	–	–

[1] Population in 1956.
[2] Calculated based upon the population in 1956 and fatal cases in 1955–1956.

Table XV. Population composition in relation to the severities of prevalence of Keshan disease (KD) in 60 villages in Beian County from 1953 to 1956

Population composition	Mild			Medium			Heavy		
	total population	died of KD	mortality (per 1,000)	total population	died of KD	mortality (per 1,000)	total population	died of KD	mortality (per 1,000)
Local residents and immigrants from near low selenium areas	8,283	22	2.6	2,491	32	12.8	1,051	19	18.1
Foreign immigrants from non-affected areas	2,416	16	6.6	1,406	48	34.2	844	85	100.7
Total	10,699	38	3.6	3,894	80	20.5	1,895	104	54.9
Mortality ratio (local/foreign)		0.39			0.37			0.18	

epidemiologic characteristics so far has not yet been satisfactorily explained. On the basis of the results obtained from early epidemiological survey in connection with recent work on selenium, Yang [1986] proposed that it is likely that people may be able to adapt to low selenium intake after long residence in a low selenium area. As far as human requirement of selenium is concerned, this metabolic adaptation should not be overlooked.

1. Difference in Mortalities between Local and Foreign Residents in KD Areas

Results from early epidemiologic surveys on Heilongjiang Province [Ju, 1956, unpublished] indicated that immigrants from non-affected areas were much more susceptible to KD. The ratio of local residents to immigrants is 1.0:2.0 in the total population while it was 1.0:6.0 in those who died of KD in 1955–1956. Later it was found, however, that among the immigrants in Beian County (neighboring Keshan County) Heilongjiang Province in northeast China, only those who immigrated from Hebei Province far apart from Beian County in north China were susceptible to KD, while those who immigrated from places near Heilongjiang actually died of KD at the same lower rate as that of local residents. Furthermore, among the fatal cases a large number of victims were those residing in the KD area for a period of 3–4 years and then the fatality rate decreased sharply as the time of residence continued for more than 10 years [Yang, J.B., 1957, unpublished]. The results are shown in table XIV. Attempts had been made to explain this fact as a phenomenon of immunity, but serum immunological studies failed to give any evidence to support this point of view.

2. The Outbreak of KD in Relation to the Constitution of Population in the Affected Areas

Studies of the population constitution of 60 villages in Beian County in relation to their severities of KD prevalence revealed the fact that the larger the proportion of foreign immigrants in the village, the heavier the outbreak of KD may appear. Data in table XV is calculated from a past report [Yang, J.B., 1957, unpublished] by Yang [1986]. Previous study in this laboratory showed that selenium levels were usually lower in areas of heavier prevalence than those of mild prevalence nearby; the results are shown in table XVI [KD Research Group of the Chinese Academy of Medical Sciences, 1979a, b].

Table XVI. Variation in hair selenium concentration (M ± SEM) of populations living in neighboring areas but with different incidences of Keshan disease

Mildly affected area		Neighboring heavily affected area	
site	hair selenium level, ppm	site	hair selenium level, ppm
Gaojian Village (1973) Xichang, Sichuan	0.112 ± 0.006	Hebian Village (1973) Mianning, Sichuan	0.071 ± 0.003
Gaojian Village (1974) Xichang, Sichuan	0.106 ± 0.007	Xiangfung Village (1974) Mianning, Sichuan	0.088 ± 0.006
Dungfung Village (1974) Chuxiung, Yunnan	0.096 ± 0.007	Shagiao Village (1974) Nanhua, Yunnan	0.068 ± 0.004
Tanjiaying Village (1974) Ansai, Shanxi	0.113 ± 0.006	Xiho Village (1974) Ansai, Shanxi	0.080 ± 0.004
Woniuho Village (1974) Puteha, Heilongjiang	0.106 ± 0.007	Zilaojiang Village (1974) Arungai, Heilongjiang	0.047 ± 0.004

The reason why the prevalence of KD became milder in recent years may be a matter of complicity. The role of selenium intervention and the general improvement of nutritional status in rural populations are of course important, but the relative stability of rural populations, particularly in the low selenium reclamation area in recent years, should not be overlooked. The stabilized population would be expected to have adapted to low selenium intake and to be able to tolerate the local poor selenium status. But, on the other hand, why did the number of KD-affected counties increase from 95 in 1959 to 206 in 1980? Probably this could be explained by the fact that since the birth of New China in 1949, grain production has tripled, but the area of cultivated land essentially remained little changed. Logically the selenium concentration of staple cereals would drop in regions near KD areas where the amount of available selenium in the soil was limited. In the newly affected areas the adaptation of children to low selenium intake may not yet have been well developed; thus, they were most susceptible to KD. The adaptation of microorganisms and animals to high selenium has already been reviewed [Levander, 1972]. Parizek et al. [1980] found that when a small dose of selenite was given 6 h before dimethylselenide, it significantly decreased dimethylselenide toxicity in the rats. Evidence shown here would indicate that adaptability of humans to low selenium may be so important as to influence the outbreak of selenium deficiency disease.

B. Selenium Requirements of Humans

Three methods of estimation have been used in this laboratory [Yang, 1985b]. Principally they are based on measuring either minimum amounts of selenium to protect inhabitants from KD by an actual dietary survey in the 'safety island' or amounts needed to maintain plasma GSHPx activity at plateau by supplementing graded amounts of seleno-*DL*-methionine to subjects in the KD area. GSHPx activity was measured by the method of Paglia and Valentine [1967]. Details of the procedures have been described elsewhere [Yang et al., 1984d]. Minimum selenium requirement was also calculated from the selenium content of marginal deficiency in staple cereals. It has been shown previously that this marginal level based on extensive analytical data is around 0.02 ppm [Yang et al., 1984c]. Since in Chinese rural populations around 70% of the dietary selenium is derived from cereals, this permitted one to calculate the daily selenium intakes for male and female adults; these values were 24 and 20 μg, respectively, with a 20% loss calculated for milling and cooking [Ferretti and Levander, 1974; Higgs et al., 1972], thus averaging 18 μg/day. The minimum selenium intakes of different age groups in the 'safety island' are shown in table XVII and the final results of adult selenium requirements obtained from different methods are summarized in table XVIII. It can be seen that the minimum selenium requirements obtained either by calculation or by actual survey are very similar, namely 16 and 18 μg/day, respectively, and averaged 17 μg/day (table XVII). This minimum requirement was considered as a critical level below which KD may occur. As adaptation and other variables are considered, this value is doubled for safety and the amount obtained is fairly comparable with the value ≤ 40 μg/day obtained for maintaining plasma GSHPx activity at a plateau (table XIV). It also agrees well with the minimum daily safe intake of 50 μg/day suggested by the US National Research Council [FNB/NRC, 1980]. In China, normal selenium intakes, excluding seleniferous areas, fell into the range of 20–160 μg/day in the majority of cases when calculated from the selenium content of

Table XVII. Daily selenium intake of different age groups residing in a 'safety island'. From Yang et al. [1984d]

Age years	Number of subjects surveyed	Intake of selenium	
		μg/day	range
1[1]	5	3.0	2.0–3.4
3–6	8	5.3	3.2–7.6
7–12	8	9.6	7.6–15.6
13–17	6	14.2	9.1–19.8
18–55	22 male	19.1	9.4–30.3
	19 female	13.3	5.4–26.1

[1] Calculated from selenium concentrations of mother's milk.

Table XVIII. Minimum and physiological selenium requirements (μg) of adults estimated by different methods. From Yang [1985b]

Method	Male	Female	Average	Minimum daily requirement of safety
Dietary survey	19.1	13.3	16	40
Calculation	19.2	16.0	18	
Plasma GSHPx activity at plateau	≤40.9	–	–	

GSHPx = Glutathione peroxidase.

staple cereals in a different province [Yang and Zhou, 1985, unpublished]. Probably a minimum selenium intake of 40 μg/day would be expected to meet the daily selenium requirement with safety in China. The exact amount of selenium needed to protect against certain human disease, such as cancer or cardiovascular disease, has not yet been settled. Levander [1983] has emphasized that before supplementation of the diet of humans with selenium can be advocated, its toxicity must be considered and proof obtained that it really acts against human cancer. In a previous paper, it was reported that inhabitants who had an average daily selenium intake of

750 μg in a site in a seleniferous area were free from pronounced clinical signs of intoxication [Yang et al., 1983]. However, in order to justify the maximum intake of safety a more detailed study is now in progress.

VII. Conclusion

Evidence has been presented that the occurrence of selenium-related endemic diseases in China depends not only upon the greatly diverse selenium levels in soil and foods, but also upon her traditional dietary habit. In this respect, rural populations subsisting largely on the grains and vegetables of local soil products is of crucial importance.

Both KD and KBD appeared only in areas with extremely poor selenium status and selenium has already been proved to be effective in the prevention of KD in many affected areas. More well-planned selenium intervention studies are needed to confirm its effect on KBD. The hypothesis arose that membrane damage caused by selenium deficiency may serve as a common basis for the pathogenesis of both KD and KBD.

For prevention of selenium deficiency, selenium intervention, was practiced in an endemic area in China. Three different methods were proposed, namely oral administration of sodium selenite tablets, enrichment of table salt with sodium selenite and foliar spraying of sodium selenite solution. Furthermore, inhabitants were encouraged to diversify their foods both in varieties and sources, and the selenium amount supplemented was kept at a minimum; it appeared that supplementation of 60–100 μg selenium daily from inorganic compounds was incapable of raising the blood selenium to the ordinary levels found in non-endemic areas at a similar selenium intake from the diet. Although it is effective for prevention purpose. Foliar spraying seems the method of choice for elevating body selenium status.

Human selenium intoxication, characterized by loss of hair and nails, has long been recognized as an endemic disease of unknown cause in south China. An extremely high selenium content has been found in the coal, soil, grain, vegetables, hair, nails and blood. The source of selenium is stony coal with an extraordinarily high selenium content.

Significantly lower cancer death rates were observed in areas with high blood selenium levels and the mortality rates of esophageal, stomach and liver cancers were also higher in the low selenium areas.

The existence of extremely low and high selenium areas in China with the occurrence of respective human endemic diseases provided a unique place for the study of selenium requirement, including tolerance limits of humans. Results obtained from studies of minimum selenium requirement and minimum safety selenium allowances are 20 and 40 μg daily, respectively. Evidence was also presented to indicate the possible existence of adaptation to low selenium intakes in inhabitants after a long residence in low selenium areas and the actual selenium intakes in the endemic area are always as low as < 10 μg daily. In view of the wide implication of selenium in human disease and health, it is proposed that the significance of selenium in human nutrition merits reappraisal.

References

Allaway, W.H.: An overview of distribution patterns of trace elements in soils and plants. Ann. N.Y. Acad. Sci. *199:* 17–25 (1972).

Bai, J.; Ge, K.; Deng, X.; Wu, S.; Wang, S.; Xue, A.; Su, C.: The effect of selenium intake on myocardial necrosis induced by viral infection in mice.

Blot, J.B.; Li, J.Y.: Some considerations in the design of a nutrition intervention trial in Linxian, People's Republic of China. Natn. Cancer Inst. Monogr. *69:* 29–34 (1985).

Bopp, B.A.; Sonders, R.C.; Kesterson, J.W.: Metabolic fate of selected selenium compounds in laboratory animals and man. Drug Metabolism Rev., vol. 13, pp. 271–318 (Dekker, New York 1982).

Brätter, P.; Virginia, E.; Negretti, S.; Rosick, U.; Jaffé, G.; et al.: Effect of selenium intake in man at high dietary levels of seleniferous areas of Venezuela. In Trace element – analytical chemistry in medicine and biology, vol. 3, pp. 29–46 (de Gruyter, Berlin 1984).

Brown, R.G.; Sweeny, R.R.; Moran, E.J., Jr: Collagen levels in tissues from selenium-deficient duck. Compar. Biochem. Physiol. *72A:* 383–389 (1982).

Chen, J.; Campbell, T.C.; Peto, R.; Li, J.: Dietary and nutritional status and cancer mortality in China. Proc. 13th Int. Nutr. Congr. 1986.

Chen, X.S.; Yang, G.Q.; Chen, J.S.; Chen, X.C.; Wen, Z.M.; Ge, K.Y.: Studies on the relations of selenium and Keshan disease. Biol. Trace Element Res. *2:* 91–107 (1980).

Chen, Y.Y.: Studies on the relationships between the epidemiological characteristics of Keshan disease and selenium status of populations and local environment in Sichuan. Acta nutr. Sin. *4:* 215–220 (1982).

Ferretti, R.J.; Levander, O.A.: Effect of milling and processing on the selenium content of grains and cereal products. Agric. Fd Chem. *22:* 1049–1051 (1974).

Fleming, C.R.; Lie, J.T.; Obrien, J.F.; Baillie, E.E.; Thistle, J.L.: Selenium deficiency and fatal cardiomyopathy in a patient on home parenteral nutrition. Gestroenterology *83:* 689–693 (1982).

FNB/NRC: Recommended Dietary Allowances; 9th ed., pp. 162–164 (Academy of Sciences, Washington 1980).
Franke, K.W.: A new toxicant occurring naturally in certain samples of plant food-stuffs. 1. Results obtained in preliminary feeding trials. J. Nutr. *8:* 597–608 (1934).
Ge, K.Y.; Bai, J.; Deng, X.J.; Wu, S.Q.; Wang, S.Q.; Xue, A.N.; Su, C.Q.: The protective effects of selenium against viral myocarditis in mice. 3rd Proc. Int. Symp. Selen. Biol. Med., Beijing 1984.
Ge, K.Y.; Chen, X.S.; Xue, A.N.; Bai, J.; Wang, S.Q.; Meng, G.S.: Morphological and histochemical observation of myocardium of Keshan disease. Acta nutr. sin. *4:* 91–95 (1982).
Ge, K.Y.; Xue, A.; Bai, J.; Wang, S.: Keshan disease – an endemic cardiomyopathy in China. Virchows Arch. Abt. A Path. Anat. *401:* 1–15 (1983).
Hadjimarkos, D.M.: Selenium in relation to dental caries. Fd Cosmet. Toxicol. *11:* 1083–1095 (1973).
Hafeman, D.G.; Suncle, R.A.; Hoekstra, W.G.: Effect of dietary selenium on erythrocyte and liver glutathione peroxidase in the rat. J. Nutr. *104:* 580–587 (1974).
Higgs, D.J.; Morris, V.C.; Levander, O.A.: Effect of cooking on selenium content of foods. Agric. Fd Chem. *20:* 678–680 (1972).
Hou, S.E.; Zhu, Z.Y.: Study of selenium status of the residents in KBD regions in China. Chin. J. Endemiol. *2:* 84–88 (1982).
Huang, Z.; Yang, Y.; Zhang, J.: The clinical signs and early diagnosis of Kashin-Beck disease. 1. Symp. of the 1982–1984 Scientific Observation on Kashin-Beck Disease in Yongshou County, pp. 122–126 (People's Medical Publishing House, Beijing 1984).
IPCS Environmental Health Criteria-Selenium: United Nations Environment Programme (ILO/WHO, Genève 1986).
Johnson, R.A.; Baker, S.S.; Fallon, J.T.; Maynard, E.P.; Ruskin, J.N.; Wen, Z.; Ge, K.Y.; Cohen, H.J.: An occidental case of cardiomyopathy and selenium deficiency. New Engl. J. Med. *304:* 1210–1212 (1981).
Keshan Disease Research Group of the Chinese Academy of Medical Sciences: Observations on the effect of sodium selenite in the preventioin of Keshan disease. Chin. med. J. *92:* 471–476 (1979a).
Keshan Disease Research Group of the Chinese Academy of Medical Sciences: Epidemiologic studies on the etiologic relationship of selenium and Keshan disease. Chin. med. J. *92:* 477–482 (1979b).
Koivistoninen, P.: Personal communication (1986).
Levander, O.A.: Metabolic interrelationships and adaptations in selenium toxicity. Ann. N. Y. Acad. Sci. *192:* 181–192 (1972).
Levander, O.A.: Recent developments in selenium nutrition; in Weininger, Briggs, Nutrition update, vol. 1, pp. 147–162 (Wiley & Sons, New York 1983).
Li, C.Z.: Observation on the efficacy of the cure of 224 KBD patients with selenite and vitamin E, and a discussion on the etiology of KBD. Chin. med. J. *59:* 169–171 (1979a).
Li, F.S.; Bai, X.F.; Qian, Y.J.; Guan, J.Y.; Duan, Y.J.; et al.: The alteration of erythrocyte plasma membrane and its significance in the pathogenesis of Kashin-Beck disease. Symp. of the 1982–1984 Scientific Observation of Kashin-Beck Disease in Yongshou County, pp. 338–344 (People's Medical Publishing House, Beijing 1984a).

Li, F.S.; Guan, J.Y.; Li, L.; Bai, C.F.; Zhao, Y.H.: The changes of selenium content and lipids composition in erythrocyte of children patients with Keshan disease and their significance in the pathogenesis of heart muscle. Acta nutr. sin. *4:* 221–226 (1982).

Li, G.: Studies on 224 cases of Kashin-Beck disease treated with Se and vitamin E, X-ray examination and etiological discussion. Natn. med. J. China *59:* 169–171 (1979b).

Li, G.: Studies on the pathogenesis of myocardial necrosis of Keshan disease (review).

Li, G.; Han, Z.; Yang, J.; Can, S.; Rin, Y.; Jia, X.; Xang, R.; Fu, G.; Lu, Z.: Studies on biochemical mechanism of the curative effect of huge intravenous dose of vitamin C on acute Keshan disease. Chin. J. Endemiol. *4:* 123–129 (1984b).

Li, G.Y.; Tsou, S.G.; Yang, T.K.; Wei, Y.Z.; Zen, Y.Y.: Determination of blood glutathione peroxidase, vitamin E, lipid peroxide and free fatty acid of children in Kashin-Beck disease – affected area. Chin. J. Endemiol. *2:* 65–68 (1983).

Li, J.Y.: An approach to the relationship between Kashin-Beck disease and selenium based upon the environmental characteristics of Shanxi Province. Huanjin Kexue *5:* 18–20 (1981).

Li, J.Y.; Li, G.Y.; Zheng, S.F.; Liu, Y.Y.; Li, P.; Yang, C.S.; Blot, W.J.; Ershow, A.G.; Li, F.P.; Creenwald, P.; Fraumeni, J.F., Jr.: A pilot vitamin intervention trial in Linxian, People's Republic of China. Natn. Cancer Inst. Monogr. *69:* 19–22 (1985a).

Li, X.; Wang, X.; Xin, S.; Jin, X.; Xuan, C.; Zhou, J.: Preliminary study on the Keshan disease-like myocardial lesion of calves. Chin. J. Endemiol. *4:* 218–221 (1985b).

Liang, S.: The prophylactic and curative effect of selenium in combatting of Kashin-Beck disease. Proc. IPCS/WPRO Meet. on Kashin-Beck disease, Beijing 1985.

Liang, S.T.: Effect of sodium selenite for prevention of Kashin-Beck disease. Symp. 1982–1984 Scientific Observation on Kashin-Beck disease in Yongshou County, pp. 142–147 (People's Medical Publishing House, Beijing 1984a).

Liang, S.T.: Observation on the sodium selenite intervention for the prevention and cure of Kashin-Beck disease. Symp. 1982–1984 Scientific Observation on Kashin-Beck Disease in Yongshou County, pp. 142–147 (People's Medical Publishing House, Beijing 1984b).

Liu, S.G.; Zhou, R.H.; Yin, T.A.; Sun, S.H.; Wang, G.Y.; Yang, G.Q.: Methods of fluorimetric determination of trace amounts of selenium in biological materials, water and soil. 2. Determination of selenium in cereals and vegetables. Acta nutr. sin. *7:* 142–147 (1985).

Lu, Y.: Studies on pathological anatomy of Keshan disease. Chin. J. Med. *48:* 6–13 (1961).

Luo, X.M.; Wei, H.J.; Yang, C.L.; Xiang, J.; Liu, X.; et al.: Bioavalability of selenium to residents in a low-selenium area of China, Proc. 3rd Int. Symp. Selen. Biol. Med., Beijing 1984.

Marco Polo: The travels of Marco Polo; translated into Chinese by Zhang Xinlang, pp. 36–96 (Commercial Press, Shanghai 1937).

McCay, P.B.; Gibson, D.D.; Fong, K.L.; Hornbrock, K.R.: Effect of glutathione peroxides acting on lipid peroxidation in biological membranes. Biochim. Biophys. Acta *431:* 459–468 (1976).

McCoy, K.E.M.; Weswig, P.H.: Some selenium responses in the rat not related to vitamin E. J. Nutr. *98:* 383–389 (1969).

Mertz, W.: Fortification of foods with vitamins and minerals. Ann. N.Y. Acad. Sci. *300:* 151–160 (1977).

Mertz, W.: Personal communication (1982).

Mo, D.: Advances in pathology, selenium and other elements research on Kashin-Beck disease. Proc. IPCS/WPRO Meet. on Kashin-Beck disease, Beijung 1985.

Mo, D.S.; Lue, Z.H.: Pathology of Kashin-Beck disease; in Zhu, Zhu, Yu, The Chinese medical encyclopaedia of edemiology, pp. 16–22 (Science & Technology Publishing Agency, Shanghai 1985).

Morris, V.C.; Levander, O.A.: Selenium content of foods. J. Nutr. *100:* 1383–1388 (1970).

Moxon, A.L.: The effect of arsenic on the toxicity of seleniferous grains. Science *88:* 81 (1938).

National Research Council (US): Subcommittee on selenium: selenium in nutrition; 2nd ed., pp. 13–19 (National Academy Press, Washington 1983).

Niu, G.; Zhang, B.; Li, J.; Zhou, Z.; Ho, S.: Two years study on selenium and Kashin-Beck disease – X-ray observation and etiological discussion. Chin. J. Endemiol. *3:* 197–201 (1984).

Noguchi, T.; Cantor, A.A.; Scott, M.L.: Mode of action of selenium and vitamin E in prevention of exudative diathesis. J. Nutr. *103:* 1502–1511 (1973).

Paglia, D.E.; Valentine, W.H.: Studies on the quantitative and qualitative characterization of erythrocyte glutathione peroxidase. J. Lab. clin. Med. *70:* 158–169 (1967).

Parizek, J.; Kalouskova, J.; Benes, J.; Parilik, Z.: Interactions of selenium – mercury and selenium – selenium compounds. Ann. N.Y. Acad. Sci. *355:* 347–360 (1980).

Piglet Illness Research Group: Preliminary study on the liver – myocardium – skeleton muscle degeneration syndrome of piglet. Heilongjiang Vet. *3:* 15–24 (1976).

Qian, P.Z.; Ge, S.; Luo, D.Y.; Wang, X.C.; Zuo, Y.Y.: Observation on the effect of selenized table salt for the prevention of Keshan disease. Chin. prev. med. J. *15:* 218–220 (1981).

Reinhold, J.G.: Trace elements – a selective survey. Clin. Chem. *21:* 276–500 (1975).

Rosenfeld, I.; Beath, O.A.: The influence of protein diets on selenium poisoning. Am. J. vet. Res. *7:* 52–56 (1946).

Rotrack, T.T.; Pope, A.L.; Ganther, H.E.; Swanson, A.B.; Hafeman, D.G.; Hoekstra, W.G.: Selenium: biochemical role as a component of glutathione peroxidase. Science *179:* 585–590 (1973).

Schrauzer, G.N.; White, D.A.; Schneider, C.J.: Cancer mortality correlation studies. 111. Statistical association with dietary selenium intakes. Bioinorg. Chem. *7:* 23–24 (1977).

Schwarz, K.; Foltz, C.M.: Selenium as an integral part of factor 3 against dietary necrotic liver degeneration. J. Am. chem. Soc. *79:* 3292–3293 (1957).

Scott, M.L.; Olson, G.; Brown, W.R.: Selenium-response myopathies of myocardium and of smooth muscle in the young poult. J. Nutr. *91:* 573–583 (1967).

Smith, M.I.: The influence of diet on the chronic toxicity of selenium. US publ. Hlth Rep. *54:* 1441–1453 (1939).

Su, C.; Gong, C.; Li, J.; Chen, C.; Zhou, D.; Jin, Q.: Preliminary results of viral etiological study of Keshan disease. Chin. med. J. *59:* 466–468 (1979).

Sun, S.Z.; Yang, G.Q.: Methods of fluorimetric determination of trace amount of selenium in biological materials, water and soil. 3. Determination of selenium in water and soil. Acta nutr. sin. *7:* 221–226 (1985).

Sun, S.Z.; Yin, T.A.; Wang, H.Z.; Yu, D.C.; Yang, G.Q.: Observation on the relationship between seasonal peak prevalence of Keshan disease and variation in hair selenium content. Chin. prev. med. J. *14:* 17–19 (1980).

Sun, S.Z.; Zhe, F.Y.; Zhou, R.H.; Yang, G.Q.: The bioavailability of soil selenium in Keshan disease and seleniferous areas. Chin. J. Endemiol. *4:* 21–28 (1985).

Sun, S.Z.; Zhou, R.H.; Gu, L.Z.; Man, R.C.; Yang, G.Q.: An investigation into the influence of dietary protein level on the selenium content and glutathione peroxidase activities in blood and tissues of rats. Acta nutr. sin. *5:* 145–150 (1983).

Tan, J.A.: The relation of natural environment to Keshan disease, the background of selenium nutrition. Acta nutr. sin. *4:* 175–182 (1982).

Tao Ping: Treatment of Keshan disease with sodium selenite (unpublished, 1978).

Thompson, C.D.; Robinson, M.F.: Selenium in human health and disease with emphasis on those aspects peculiar to New Zealand. Am. J. clin. Nutr. *33:* 303–323 (1980).

Thompson, J.N.; Scott, M.L.: Improved lipid and vitamin E absorption related to atrophy of the pancrease in selenium-deficient chicks. J. nutr. *100:* 797–809 (1970).

Underwood, E.J.: Selenium in trace elements in human and animal nutrition; 4th ed., pp. 302–340 (Academic Press, New York 1977).

Veterinary Station of Jiusan Farm: Experimental protection of liver and muscle necrosis of piglet with selenium and vitamin E. Heilongjiang Vet. *3:* 23–26 (1976).

Wang, F.: Discussion on the etiology of Keshan disease based upon the pathological findings. Chin. med. J. *48:* 17–21 (1962).

Wang, G.A.; Zhou, R.H.; Sun, S.Z.; Yin, T.A.; Yang, G.Q.: Differences between blood selenium concentration of residents in Keshan disease-affected and non-affected areas. Correlation between selenium content of blood and hair. Chin. prev. med. J. *13:* 204–206 (1979).

Wang, G.Y.; Zhou, R.H.; Sun, S.H.; Yin, T.A.; Liu, S.G.: Yang, G.Q.: Methods of fluorometric determination of trace amount of selenium in biological materials, water and soil. 1. Determination of selenium in hair, blood and urine. Acta ntur. sin. *7:* 39–45 (1985).

Wang, Y.; Zhu, C.; Ying, M.: A correlative study on the X-ray and pathological changes in Kashin-Beck's disease. Natn. med. J. China *64:* 357–359 (1984a).

Wang, Y; Zhu, C.; Ying, M.; Ren, H.; Mao, D.; Xeng, S.; Liu, Y.: The pathology and evolution of the X-ray signs of the bone and cartilage in Kashin-Beck's disease. Chin. J. Radiol. *18:* 81–84 (1984b).

Wang Zhiwu: Personal communication (1972).

Watkinson, J.H.: Fluorometric determination of selenium in biological material with 2,3-diamino-naphthalene. Analyt. Chem. *38:* 92–97 (1966).

Wen, Z.; Chen, X.; Fu, P.; Qian, P.; Liu, R.; Huang, J.: Keshan disease survey in a population treated with long-term selenium supplementation. Chin. J. Endemiol. *4:* 73–76 (1985).

Wo, W.; Yang, F.Y.: Reasearch on selenium and erythrocyte membrane. 11. The protective effect of selenium to erythrocyte. Zhongguo Kexue, Ser. B *4:* 401–406 (1986).

Wu, A.S.H.; Oldfield, J.E.; Shull, L.R.; Cheeke, R.R.: Specific effect of selenium deficiency on rat sperm. Biol. Reprod. *20:* 793–798 (1973).

Xian Medical College: Observation on the effects of sodium selenite for prevention of acute Keshan disease. Chin. med. J. *59:* 407–460 (1979).

Xichang County Antiepidemic Station: Five years prevention of Keshan disease with selenium (unpublished).

Xu, G.L.: Selenium status and dietary selenium content of a population in the endemic and non-endemic areas of Keshan disease. Acta nutr. sin. *4:* 183–190 (1982).

Yang, F.Y.; Huang, F.; Lin, Z.; Wo, W.: Abnormalities of erythrocyte membrane from patients with Kashin-Beck disease. Proc. 3rd Int. Symp. Selen. Biol. Med., Beijing 1984a.

Yang, F.Y.; Huang, F.; Lin, Z.H.; Zhang, K.; et al.: The biochemical alteration of erythrocyte plasma membrane of Kashin-Beck disease patients in Yongshou. Symp. 1982–1984 Scientific Observation on Kashin-Beck Disease in Yonghsou County, pp. 366–371 (People's Medical Publishing House, Beijing 1984b).

Yang, G.Q.: Research on selenium-related problems in human health in China. Proc. 3rd Int. Symp. Biol. Med., Beijing 1984a.

Yang, G.Q.: Keshan disease: an endemic selenium-related deficiency disease; in Chandra, Trace Elem. Nutr. Childr. Nestlé Nutr. vol. 8, pp. 273–290 (Raven Press, New York 1985a).

Yang, G.Q.: Keshan disease, selenium deficiency and adaptation. Proc. 13th Int. Conf. Nutr., Brighton 1985b.

Yang, G.Q.: Comment on the recent epidemiologic characteristics in relation to the possible existence of adaptation to low selenium intakes in inhabitants after long residence in a Keshan disease area. Wesheng Yanju *15:* 23–27 (1986).

Yang, G.Q.; Chen, J.S.; Wen, Z.M.; Ge, K.Y.; Zhu, L.Z.; Chen, X.C.; Chen, X.S.: The role of selenium in Keshan disease; in Drapper, Advances in nutrition research, vol. 6, pp. 203–231 (Plenum Press, New York 1984c).

Yang, G.Q.; Wang, G.Y.; Yin, T.A.; et al.: Relation between the distribution of Keshan disease and selenium status. Acta nutr. sin. *4:* 191–200 (1982a).

Yang, G.Q.; Wang, S.Z.; Zhou, R.H.; Sun, S.Z.; et al.: Investigation on loss of hair and nail disease of unknown etiology – endemic selenosis. Acta Acad. Med. Sin. (suppl. 2) *3:* 1–6 (1981).

Yang, G.Q.; Wang, S.Z.; Zhou, R.H.; Zhou, R.H.; Sun, S.Z.: Endemic selenium intoxication of humans in China. Am. J. clin. Nutr. *37:* 872–881 (1983).

Yang, G.Q.; Yin, T.A.; Lue, S.J.; Wang, G.A.; Chow, Z.A.; et al.: Approaches to the supplementation of selenium in the prevention of Keshan disease. Acta nutr. sin. *4:* Ho (1982b).

Yang, G.Q.; Yin, T.A.: Sun, S.Z.; Yu, D.C.: Hair selenium levels of vulnerable populations of Keshan disease. Chin. prev. med. J. *14:* 14–16 (1980).

Yang, G.Q.; Zhu, L.Z.; Liu, S.J.; Gu, L.Z.; Huang, T.H.; Lu, M.D.: Studies of human requirements in China. Proc. 3rd Int. Symp. Biol. Med., Beijing 1984d.

Yang, T.S.: Reduction of activities of respiratory enzymes in the heart tissues of Keshan disease patient. Difangbing Tongxun *3:* 1–3 (1982).

Yang, T.S.: Advances in the biochemistry of Kashin-Beck disease. Symp. 1982–1984 Scientific Observation on Kashin-Beck Disease in Yongshou County, p. 225–286 (People's Medical Publishing House, Beijing 1984b).

Yang, Y.Y.; Du, B.Z.: Causal relationship of environmental selenium levels to Keshan disease. Difangbing Tongxun *1:* 57–60 (1983).

Yin, P.: The concept of Kashin-Beck disease and its diagnosis. Proc. IPCS/WPRO Meet. on Kashin-Beck Disease, Beijing 1985.

Yin, P.; Guo, X.; Zhang, S.; Bai, C.: A comparative study of comprehensive method and Se-fortified wheat for prevention and therapy of Kashin-Beck disease. Proc. IPCS/WPRO Meet. on Kashin-Beck disease, Beijing 1985.

Yin, S.A.; Gu, L.Z.; Zhou, R.H.; Yang, G.Q.: Comment on the daily selenium intake of staff and children of nursery and kindergarden in Beijing. Acta nutr. sin. *7:* 195–701 (1985).

Yin, T.A.; Su, S.Z.; Wang, S.Z.; You, D.Q.; Yang, G.Q.: Difference of the amounts of selenium excretion in urine in children of Keshan disease-affected and non-affected areas. Chin. prev. med. J. *132:* 207–210 (1979).

Yu, S.Y.; Chu, Y.J.; Gong, X.L.; Hou, C.: Reginal variation of cancer mortality incidence and its relation to selenium levels in China. Biol. Trace Elem. Res. *7:* 21–29 (1985).

Yu, S.Y.; Wang, L.M.: Different effects of selenium on cyclic AMP metabolism in hepatoma cells and normal liver cells. Biol. Trace Elem. Res. *5:* 9–16 (1983).

Yu, S.Y.; Zhao, Z.C.; Zhu, Y.J.: Studies on the action of alphatocopherol and selenium on the serum lipid peroxide of selenium on the function and structure of mitochondria isolated from hepatoma and normal liver. Proc. 3rd Int. Symp. Selen. Biol. Med., Beijing 1984a.

Yu, S.Y.; Zhu, Y.J.; Hou, C.; Huang, C.: Selective effects of tumor-bearing rat (in Chinese). Chin. J. Cancer *6:* 5–7 (1984b).

Yuan, C.H.: A discussion on the etiology of Keshan disease from the epidemiologic point of view. Difangbing Tongxun *47:* 35–38 (1983).

Zhou, R.H.; Sun, S.Z.; Zhai, F.Y.; Man, R.E.; Yang, G.Q.: Influence of dietary protein level on the selenium contents and glutathione peroxidase activities of blood and tissues of rats. Acta nutr. sin. *5:* 137–144 (1983).

Zhu, Y.W.; Endemic disease; In Zhu, Zhu, Yu, 1983 Chinese Cyclopaedia: endemiology, p.1 (Science & Technology Publishing Agency, Shanghai 1983).

Zhu, L.Z.; Xia, Y.M.; Yang, G.Q.: Blood glutathione peroxidase activities of populations in Keshan disease-affected and non-affected areas. Acta nutr. sin. *4:* 229–233 (1982).

Guangqi Yang, MD, Institute of Nutrition and Food Hygiene,
Chinese Academy of Preventive Medicine, 29 Nan Wei Road, Beijing 2 (China)

Wld Rev. Nutr. Diet., vol. 55, pp. 153–164 (Karger, Basel 1988)

Possible Nutrient Mediators in Psoriasis and Seborrheic Dermatitis

I. Prevalence, Etiology, Symptomatology, Histological and Biochemical Features

Sarah Brenner, Crystal Horwitz

Department of Dermatology, Ichilov Hospital, Tel Aviv, and
Department of Gastroenterology, Ichilov Hospital, Tel Aviv, Israel

Contents

Psoriasis

Epidemiology and Prevalence

Psoriasis is an inflammatory disease of the skin, mainly of the elbows, knees and scalp, characterized by silvery scales on reddish plaques. It may be benign, acute or chronic. Serious, life-threatening forms are known to occur.

In 1978 the United States Department of Health, Education and Welfare published detailed estimates of the frequency of psoriasis as part of a survey of skin disease in the population [1]. Of 20,749 people examined, 0.5% were found to have psoriasis. However, a prevalence somewhat higher than this has been cited in the range 2–3% [2] and Lomholt [3] found that the prevalence in the Faroe Islands was 2.8%.

Psoriasis has been reported to affect all racial groups, is more prevalent in adults than children and equally distributed with regard to gender. It is rare among the North and South American Indians; no psoriasis has been found among the natives of American Samoa and in certain regions of the South American Andes [4]; it is virtually unknown among the Eskimos [ref. 119, part II].

Clinical Appearance

The typical lesion in psoriasis appears as rounded and circumscribed red patches covered by silvery-white scales. They are most frequently located on the elbows, knees, scalp and sacral region, usually in a symmetrical distribution. The nails have a characteristic stippled appearance. The Auspitz sign is diagnostic of psoriasis: upon removal of scale from a psoriatic lesion, bleeding points appear. The Koebner phenomenon is also typical. This is an isomorphic response appearing at sites of injury, for example at scratches or scars. This phenomenon is non-specific and can occur in other skin disorders. Itching or burning sensations may occur at sites of lesions (see fig. 1).

These symptoms are classical for psoriasis vulgaris. Other forms are: (1) psoriasis guttata, in which the lesions appear as drops (gutta = drop) all over the body. These are erythematous and covered by silvery scales. Psoriasis guttata may occur after an infection, particularly streptococcal. (2) Psoriasis inversa or psoriasis plexura, found in intertriginous areas, such as the axilla, submammary folds, pubis, groin and gluteal crease. In these conditions the plaques are less scaly, more moist and often exhibit fissuring. (3) Generalized pustular psoriasis is the severest form and can be fatal. Pustules appear abruptly, sometimes coalescing to form 'lakes' of pus. Intense itching occurs, fever and malaise may be present. This form may involve mucous membranes as well. Iodides and salicylates have been known to provoke this form. (4) Psoriasis palmoplantaris, a pustular form involving mainly the soles and palms, and (5) erythrodermic psoriasis (EP) a severe form which could be an initial form of another type of psoriasis, or an aggravation of an existing, widespread psoriasis. Systemic complica-

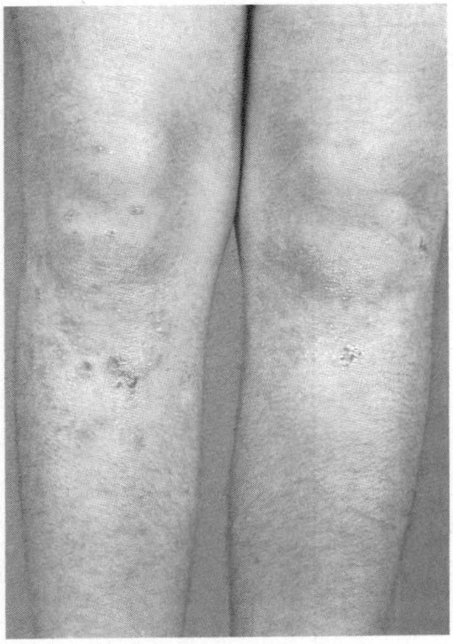

Fig. 1. Psoriasis vulgaris.

tions or fever may be present. Periods of remission and exacerbation characterize most forms of psoriasis. A form of arthritis accompanies psoriasis in about 15–20% of patients. Characteristic of the condition is the inflammatory involvement of the terminal interphalangeal and sacroiliac joints. Typical deformations of bone appear and typical roentgenological signs are present. HLA B-27 is present in about 41% of patients with psoriatic arthritis. Rheumatoid factor is absent and rheumatic nodules are not seen. Uric acid levels are high in psoriatic arthritis (possibly reflecting the rapid skin turnover), but the incidence of gout is not higher than in the normal population [5]. Psoriasis usually precedes the onset of psoriatic arthritis, but during an attack the psoriatic lesions may be absent or minute and virtually undetectable, especially on the scalp. Nail pitting or onycholysis, almost always present in psoriasis, is then used as a diagnostic aid.

Etiology

The etiology of psoriasis is still unknown, but certain factors can be considered for further understanding of the disease. These are:

Pathogenesis. In psoriasis the mitotic index (number of cell divisions per 100 cells per unit of time) is increased. The germinative cell population and the epidermo-dermal junction surface areas increase. There is a proliferation, dilatation and tortuosity of the capillaries of the dermal papillae and a migration of polymorphonuclear leukocytes into the epidermis. Epidermal turnover time is increased from the normal 25–30 days to 3–4 days [6] (see Histopathology).

Provocation. It may be more accurate to use the phrase 'influencing factors', since it is not known with surety that provocation actually triggers an attack. These factors are:

Stress. Anxiety, depression or personal problems can precipitate or exacerbate psoriasis in 23–42% of patients [7–9]. In one study 80% of patients believed that worry was associated with a flare of disease activity [10]. Farber et al. [11] raised the role of stress not only as a factor in provoking or exacerbating an attack, but also as a determinant of the distribution of plaques. They suggested a role for neuropeptides in the etiopathogenesis of psoriasis and postulated that substance P was a modulator of the inflammatory response.

Infection. After throat infections caused by β-hemolytic streptococci psoriasis can flare up, especially in children [12a]. In these cases psoriasis is frequently of the guttate type. Vaccination or upper respiratory tract infections in adults occasionally exacerbate the disease.

Drugs. β-Blockers [12b], lithium carbonate [13] and α-interferon [14] have been shown to provoke or exacerbate psoriasis. Mechanisms are obscure but may involve the epidermal leukocytes directly, influence efficacy of antipsoriatic therapy or change immune status.

Trauma. After mechanical injury, psoriasis can appear at the site of the trauma after a latent period of about 14 days. This Koebner phenomenon (see above) is well known and it can be elicited in other skin diseases or fungal infections. Ultraviolet (UV) light, heat and low humidity are other physical factors that can provoke psoriasis.

Genetic Predisposition. A familial pattern of psoriasis frequently occurs [3, 7, 15, 16]. HLA frequency studies in psoriasis have shown associations with antigens associated with the BC and D loci. HLA B-13 and B-17 were found to be significantly increased in psoriasis [17] and HLA Cw6 has the most significant association with the disease, 87% of patients with psoriasis vulgaris possess this antigen compared to 33% of controls [18].

Since psoriasis is not a homogeneous disease and has different clinical manifestations, HLA typing has been done in different population groups and correlated with age of onset and severity and type of disease. HLA B-17, B-37 and Cw6 types tend to have psoriasis at an early age [19–21] while those with Bw16 have a late age of onset [22]. HLA B-16, B-17 and B-37 types tend to have a more severe psoriasis [22], while guttate psoriasis as well as psoriasis vulgaris has been associated with Cw6. Generalized pustular psoriasis is associated with HLA B-27 [23], while erythrodermic psoriasis is associated with HLA A-2, A-28, B-13 and B-17 [20]. HLA B-27 is found in 41% of patients with psoriatic arthritis [24]. The data of HLA antigen frequency are as yet inconclusive, but it is possible that one or more HLA-linked genes may play a role as determinants of the etiopathogenesis of psoriasis.

Psoriasis probably belongs to the diseases with multifactorial inheritance, in which the genetic components are alleles at an unspecified number of loci, each contributing a small but significant effect. They interact with environmental factors to cause the appearance of lesions or traits in question. Patients with psoriasis may be thought of as succumbing to a critical combination of causative genetic and environmental factors.

Histochemistry

An increase in mast cells in psoriatic skin has been reported; abnormal granules in these cells may represent partial degranulation [25]. The mast cell metachromatic granules contain several mediators of inflammation including histamine, heparin, prostaglandins (PG), leukotrienes and proteinases [26].

The release of histamine may be responsible for the pruritis that accompanies psoriasis [27]. Histamine has also been implicated in epidermal cell function [28, 29]. The erythema, vascular dilatation, permeability and keratinocyte proliferation seen [30] may be due to mast cell degranulation and release of histamine and heparin [31, 32].

The release of mast cell and macrophage arachidonate metabolites such as prostaglandins and leukotrienes may contribute to the early inflammation that is seen in psoriatic lesions and that precedes keratinocyte proliferation [33] but their role and influence has not been conclusively determined.

Mast cell and leukocyte neutral proteinases have also been thought to play an important part in the pathogenesis of psoriasis, both as inflammatory mediators [26, 34] and by their ability to alter the keratinocyte cell surface [35, 36]. Lymphocytes are the first inflammatory cells to appear at psoriatic sites [37]. Their activation results in the production of various soluble lymphokines, including interferons and interleukins. These may alter mitotic activity [38]. It has been shown that α- and γ-interferons inhibit keratinocyte proliferation in tissue culture [39], but their role in the provocation of psoriasis is not clear [14].

Voorhees and Duell [40] proposed that the pathogenesis of psoriasis could be explained by a defective adenosine 3′:5′-cyclic phosphate (cAMP) system, since they found low epidermal cAMP and adenylcyclase activity in psoriatic epidermis. Aso et al. [41] reported a reduced ability of involved epidermal areas in psoriasis to produce PGE and that this reduced ability might account for the low levels of cAMP in the lesions.

Several conflicting reports followed. Using an improved skin biopsy technique to avoid any artificial rise in cAMP due to ischemia, Yoshikawa et al. [42] and Adachi et al. [43] determined the cAMP levels in psoriatic lesions. They concluded that cAMP levels in the involved epidermis was not lower than that in the uninvolved epidermis of psoriatic patients.

The first histological change in psoriasis is an infiltration into the epidermis by neutrophilic leukocytes [44]. Tagami and Ofuji [45] found a chemotactic substance in psoriatic scales by a modified Boyden's chamber technique and postulated that it might be a complement-derived chemotactic factor. Recently Brain et al. [46] reported a release of leukotriene B_4 (LTB_4) from abraded psoriatic plaque. They have also shown that cultured epidermal cells release small amounts of LTB_4-like material [47]. These data suggest that some chemotactic factors may play a role in the pathogenesis of psoriasis (fig. 1).

Skin lesions of psoriatic patients have a higher arachidonic acid concentration than normal skin of the same patients [48]. Recently LTB_4 has been found in psoriatic epidermis [46], probably formed by keratinocytes [47]. Oral benoxaprofen causes amelioration of psoriatic lesions [49]. It is assumed that the mechanism of action is either by the inhibition of epidermal 5-lipoxygenase or by inhibiting indirectly the accumulation of phagocytes in psoriatic lesions [50, 51]. A pathogenic role in the etiology of psoriasis for arachidonic acid and its metabolites has recently been reviewed [52].

Histopathology

The characteristic histopathological features seen in psoriasis are: (a) regular elongation of the rete ridges with thickening in their lower levels; (b) elongation and edema of the papillae; (c) relative thinning of the suprapapillary portions of the stratum malpighii with the occasional presence of very small 'spongiform pustules'; (d) absence of granular cells; (e) parakeratosis; (f) presence of Munro microabscesses.

No edema, neither inter- nor intracellular, is seen in the rete ridges. The papillae show edema and the capillaries in the upper portions of the papillae are dilated. A mild to moderate inflammatory infiltrate is present in the upper dermis; the papillae, which consist of mononuclear cells (except in early lesions), contain neutrophils in their upper portions. The suprapapillary plate is relatively thin. The parakeratotic stratum corneum may be infiltrated by neutrophils forming spongiform pustules of Kogoj. These are diagnostic for psoriasis. The stratum granulosum may be entirely absent and cases occur in which there coexist areas with and without it.

The size and number of keratohyaline granules are greatly reduced and occasionally they are absent. Since an inverse relationship exists between the absence of keratohyaline granules and the presence of parakeratosis, it is important to remember that the horny layer in some cases consists entirely of parakeratotic cells.

The Munro microabscesses are located within parakeratotic areas of the horny layer. They consist of an accumulation of pyknotic neutrophils that have migrated from the capillaries in the papillae through the suprapapillary epidermis.

The bleeding points (Auspitz sign) that are produced after removing the scale are due to the parakeratosis, to the intercellular edema of the keratinocytes in the suprapapillary epidermis and to the dilatation of the capillaries in the upper portion of the papillae.

Treatment

Treatment for psoriasis may be either topical or systemic.

Topical treatment

Corticosteroids. These are administered in various forms (ointments in aqueous or fatty bases, solutions) according to severity and areas affected. The scalp for example, would be treated with a solution. In addition to these simple techniques of application, occlusive dressing may be used.

Anthralin. Topical application of ointments containing 0.1–1% and more has been used. Occasionally pigmentation occurs, especially in scaly psoriasis, but this disappears.

Tars. Soaps, shampoos and ointments and lotions, sometimes in combination with corticosteroids or salicyclic acid are used.

Salicylic Acid. This is used in combination with any of the above agents in 3–5% concentrations.

Goeckerman Technique. A 2–5% tar preparation is applied or a tar bath taken and UV light treatment is given 24 h later.

Ultraviolet Light. Natural sun baths improve psoriasis. Especially beneficial is the location of the Dead Sea. In addition to natural sources of UV light, artificial long wave (320–480 nm) UV lamps are used. Care must be taken to avoiding burning of the skin and in increasing the UV dosage. The Koebner phenomenon may be provoked.

Systemic Treatment

PUVA Treatment. A combination of psoralen taken orally and UV irradiation (UVA) 2 h later causes a phototoxic reaction that is beneficial. Long wave UV is obtained from sun lamps.

Methotraxate. This cytotoxic, folic acid antagonist is used to retard the epidermal cell turnover rate by blocking the synthesis of DNA. Methotraxate can be given orally, intramuscularly or intravenously, according to requirements.

Retinoic Acid. In severe cases aromatic retinoids are used with considerable benefit. Unfortunately, side effects can be severe, and include cheilitis, xerosis, facial dermatitis and hair loss. The drug is teratogenic and should not be given in pregnancy.

Peritoneal Dialysis. This and hemodialysis have been effective in the treatment of psoriasis but are still considered heroic methods.

Seborrheic Dermatitis

Epidemiology and Prevalence

Seborrheic dermatitis is thought to afflict 2–5% of adolescent and adult populations. Men and women are equally affected and it knows no racial boundaries.

Clinical Appearance

SD is a chronic or acute papulosquamous inflammatory disease of the skin, primarily of areas richly supplied with sebaceous glands (hypersecretion of sebum is not, however, necessarily associated with SD). These are the scalp, eyebrows, eyelids, nasolabial folds, aural canals, auditory meatus, retroauricular regions, axillae, sternal areas, interscapular areas, umbilicus and body folds. The most prominent feature of the condition is the appearance of scale, which can present as dry (sicca type) or oily. SD occurs most frequently on the scalp where the scale is dry, snow-white and desquamates (dandruff). A similar condition occurs in newborn infants, called cradle-cap, but it is not known whether the etiology is similar. Lesions may be accompanied by erythema, exudation and pruritus. Secondary infections may develop, accompanied by impetigo. SD may progress to a generalized exfoliative dermatitis.

Etiology and Pathogenesis

The etiology of SD is unknown. It may be based upon a genetic predisposition, with aggravating or precipitating factors, such as emotional stress, climatic conditions, nutritional status and hormonal status. It is as yet not certain what role, if any, the fungus *Pityrosporum ovale* plays in SD. It may be an opportunistic infection without influence on the outcome of the condition, or it may play a more significant role in its pathogenesis. SD sometimes is found concomitant with certain organic diseases, such as Parkinson's disease, diabetes mellitus, malabsorption syndromes and epilepsy. Neuroleptic drugs can precipitate SD [53]. Patients have periods of remission and exacerbation, and a tendency to lifelong recurrences exists. Outbreaks may last for weeks, months or years.

Histopathology

SD could be considered a condition histologically half way between psoriasis and chronic dermatitis/eczema. Parakeratosis is seen in focal areas of the horny layer cells, similar to those in Munro miscroabscesses of psoriasis. Slight to moderate acanthosis is found with some elongation of the rete ridges, similar to the histopathology of psoriasis. Slight intracellular edema and spongiosis similar to that found in eczema is seen and the dermis has a mild, chronic inflammatory mixed-cell infiltrate, found perivascularly. Neutrophils migrate into the epidermis.

Treatment

Corticosteroids, antibiotics and topical zinc/tar/selenium preparations are the cornerstones of treatment. Recently metronidazole (an antiprotozoal drug) has been successfully used [54] and lithium succinate applied topically has also been successful [55]. Agents that reduce the amount of sebum produced, such as 13-*cis*-retinoic acid and PUVA treatment have not been satisfactory [56]. Physicians occasionally prescribe vitamin preparations as lesser adjuncts.

References

1 United States Department of Health, Education and Welfare: Skin conditions and related needs for medical care among persons 1–74, United States 1971–74, US Department of Health. Education and Welfare Vital and Health Statistics Series 11, No. 212, November 1978.

2 Johnson, M.L.; Stern, R.S.: Prevalence of skin disorders: in Fitzpatrick et al., Dermatology in general medicine; ed. 2, pp. 3–7 (McGraw-Hill, New York 1979).

3 Lomholt, G.: Prevalence of skin diseases in a population; a census study from the Faroe Islands. Dan. med. Bull. *II:* 1–7 (1964).

4 Convit, J.: Investigation of the incidence of psoriasis among Latin American Indianers. Proc. 12th Int. Congr. of Dermatology, Washington 1962. Int. Congr. Ser. No. 55 Amsterdam (1963).

5 Bird, H.A.; Wright, V.: Psoriatic arthritis. Clin. rheum. Dis. *9:* 671–680 (1983).

6 Weinstein, G.D.; Van Scott, E.J.: Autoradiographic analysis of turnover times of normal and psoriatic epidermis. J. invest. Derm. *45:* 257–262 (1965).

7 Farber, E.M.; Nall, M.C.: The natural history of psoriasis in 5,600 patients. Dermatologica *148:* 1–18 (1974).

8 Newbold, P.C.H.: Pruritus in psoriasis; in Farber, Cox, Psoriasis. Proc. 2nd Int. Symp. on Psoriasis, New York 1977, pp. 334–336 (Yorke Medical Books, New York 1977).

9 Brown-Falco, O.; Burg, G.; Farber, E.M.: Psoriasis: eine Fragenbogenstudie bei 536 Patienten. Münch. med. Wschr. *114:* 1–15 (1972).

10 Gilbert, A.R.; Dodgers, D.A.; Roenigk, H.H.: Personality evaluation in psoriasis. Cleve. clin. Q. *40:* 147 (1973).

11 Farber, E.M.; Nickoloff, B.J.; Recht, B.; Fraki, J.E.: Stress symmetry and psoriasis: possible role of neuropeptide. J. Am. Acad. Derm. *14:* 305–311 (1986).

12a Asboe-Hensen, G.: Psoriasis in childhood; in Farber, Cox, Psoriasis. Proc. 1st Int. Symp. on Psoriasis (University of California Press, Stanford 1971).

12b Arntzen, N.; Kavli, G.; Volden, G.: Psoriasis provoked by beta-blocking agents. Acta derm.-vener., Stockh. *64:* 346–348 (1984).

13 Skoven, S.; Thorman, J.: Lithium compound treatment and psoriasis. Archs Derm. *115:* 1185–1187 (1979).

14 Quesada, J.R.; Gutterman, J.U.: Psoriasis and alpha-interferon. Lancet *i:* 1466–1468 (1986).

15 Hellgrenn, L.: Psoriasis, the prevalence in sex and age and occupational groups in total populations in Sweden: morphology, inheritance and association with other skin and rheumatic diseases, Stockholm 1967.

16 Watson, W.; Cann, H.M.; Farber, E.M.: The genetics of psoriasis. Archs Derm. *105:* 197–207 (1972).

17 Russel, T.J.; Schultz, L.M.; Kuban, D.J.: Histocompatibility (HLA) antigens associated with psoriasis. New Engl. J. Med. *287:* 738–740 (1972).

18 Gazit, E.; Brenner, S.; Efter, T.; Orgad, S.; Mizrachi, Y.; Krakowski, A.: HLA in psoriasis vulgaris. Tissue Antigens *10:* 206 (1977).

19 Brenner, W.; Gschnait, F.; Mayr, W.R.: HLA B-13, B-17, B-37 and CW-6 in psoriasis vulgaris association with age of onset. Archs Derm. Res. *262:* 337–339 (1978).

20 Karvonen, J.: HLA antigens in psoriasis with special reference to the clinical type, age of onset, exacerbation after respiratory infection and occurrence of arthritis. Ann. clin. Res. *7:* 301–311 (1975).
21 Svejgaard, A.; Staub Nielsen, L.; Svejgaard, E.: HLA in psoriasis population and family studies. Br. J. Derm. *91:* 145–153 (1974).
22 Krulis, L.; Farber, E.M.; Grumet, F.C.: Histocompatibility (HLA) antigen in psoriasis. Archs Derm. *111:* 857–860 (1975).
23 Zachariae, H.; Overgaard-Petersen, H.; Kissmayer-Nielsen, F.; et al.: HLA antigen in pustular psoriasis. Dermatologica *154:* 73–77 (1977).
24 Zachariae, H.; Hjortshøoj, A.; Kissmayer-Nielsen, F.; et al.: HLA antigens and psoriatic arthritis. Acta derm.-vener., Stockh. *54:* 443–447 (1974).
25 Cox, A.J.: Mast cells in psoriasis; in Farber, Cox, Psoriasis. Proc. 2nd Int. Symp. New York 1977, pp. 34–43 (Yorke Medical Books, New York 1977).
26 Fraki, J.E.; Schechter, N.M.; Lazarus, G.S.: Human skin proteinases as inflammatory mediators. Br. J. Derm. *1095:* 72–76 (1983).
27 Newbold, P.C.H.: Pruritus in psoriasis; in Farber, Cox, Psoriasis. Proc. 2nd Int. Symp., New York 1977, pp. 334–336 (Yorke Medical Books, New York 1977).
28 Harper, R.A.; Flaxman, B.: Effects of pharmaceutical agents on human keratinocytes, migration in vitro III. Inhibition by histamine and methylated analogs. J. invest. Derm. *65:* 400–403 (1975).
29 Aoyagi, T.; Adachi, K.; Halprin, K.M.; et al.: The effect of histamine on epidermal outgrowth: its possible dual role as an inhibitor and stimulator. J. invest. Derm. *76:* 24–27 (1981).
30 Braverman, I.M.; Sibley, J.: Role of the microcirculation in the treatment and pathogenesis of psoriasis. J. invest. Derm. *78:* 12–17 (1982).
31 Wolf, J.E.; Hubler, W.R.: Angiogenesis in psoriasis; in Farber, Cox, Psoriasis. Proc. 2nd Int. Symp., New York 1977, pp. 375–377 (Yorke Medical Books, New York 1977).
32 Brody, I.: Mast cell degranulation in the evaluation of acute eruptive guttate psoriasis vulgaris. J. invest. Derm. *82:* 460–464 (1984).
33 Brown-Falco, O.; Christophers, E.: Structural aspects of initial psoriatic lesions. Archs Derm. Res. *251:* 95–110 (1974).
34 Schwartz, L.B.; Austen, K.F.: Structure and function of the clinical mediators of mast cells. Prog. Allergy *34:* 271–321 (1984).
35 Fraki, J.E.; Jakoi, L.; Davis, A.O.; et al.: Polymorphonuclear leukocyte function in psoriasis: chemotaxis, chemokinesis, betraadrenergic receptors and proteolytic enzymes of polymorphonuclear leukocytes in peripheral blood from psoriatic patients. J. invest. Derm. *81:* 254–257 (1983).
36 Orfanos, C.E.; Schaumburg-Lever, G.; Mahrle, G.; Lerner, W.F.: Alteration of cell surface as a pathogenic factor in psoriasis: possible loss of contact inhibition of growth. Archs Derm. *107:* 38–46 (1973).
37 Ragaz, A.; Ackerman, A.B.: Evolution, maturation and repression of lesions of psoriasis. Am. J. Dermatopathol. *1:* 199–214 (1979).
38 Korszum, A.K.; Wilton, J.M.; Johnson, W.W.: The in vivo effect of lymphokines on mitotic activity and keratinization in guinea pig epidermis. J. invest. Derm. *76:* 433–437 (1981).

39 Nickoloff, B.J.; Bosham, T.Y.; Merigan, T.C.; Morhenn, V.B.: Antiproliferative effects of recombinant alpha and gamma interferon on cultured human keratinocytes. Lab. Invest. *51:* 697–701 (1984).
40 Voorhees, J.J.; Duell, E.A.: Psoriasis as a possible defect of the adenylcyclase cyclic AMP cascade: a defective chalone mechanism? Archs Derm. *104:* 352–358 (1971).
41 Aso, K.; Deneau, D.G.; Krulig, L.; Wilkinson, D.I.; Farber, E.M.: Epidermal synthesis of prostaglandins and their effect on levels of cyclic/adenosine 3′5′ monophosphate. J. invest. Derm. *64:* 326–331 (1975).
42 Yoshikawa, K.; Adachi, K.; Halprin, K.M.; Levine, V.: Is the cyclic AMP in psoriatic epidermis low? Br. J. Derm. *93:* 253–258 (1975).
43 Adachi, K.; Iizuka, H.; Halprin, K.M.; Levine, V.: Epidermal cyclic AMP is not decreased in psoriasis lesions. J. invest. Derm. *74:* 74–76 (1980).
44 Baker, H.; Wilkinson, D.S.: Psoriasis; in Book, Wilkinson, Ebling, Textbook of dermatology, pp. 1315–1367 (Blackwell, Oxford 1979).
45 Tagami, H.; Ofuji, S.: Leukotactic properties of soluble substances in psoriatic scale. Br. J. Derm. *95:* 1–8 (1976).
46 Brain, S.D.; Camp, R.D.R.; Dowd, O.M.; et al.: Psoriasis and leukotriene B_4. Lancet *ii:* 762–763 (1982).
47 Brain, S.D.; Camp, R.D.R.; Leigh, I.M.; Ford-Hutchinson, A.W.: The synthesis of leukotrine B_4-like material by cultured human keratinocytes. J. invest. Derm. *78:* 328 (1982).
48 Hammarstrom, S.; Hamberg, M.; Samuelsson, B.; Duell, E.A.; Stawinski, M.; Voorhees, J.J.: Increased concentrations of non-esterified arachidonic acid, 12L-hidroxy-5,8,10,14-eicosatetraenoic acid, prostaglandin E_2 and prostaglandin F_2-alpha in epidermis of psoriasis. Proc. natn. Acad. Sci. USA *72:* 5130–5134 (1975).
49 Allen, B.R.; Littlewood, S.M.: Benoxaprofen effect on cutaneous lesions in psoriasis. Br. med. J. *285:* 1241 (1982).
50 Kragballe, K.; Herlin, T.: Benoxaprofen improves psoriasis. Archs Derm. *119:* 548–552 (1983).
51 Voorhees, J.J.: Leukotrienes and other lipoxygenase products in the pathogenesis and therapy of psoriasis and other dermatoses. Archs Derm. *119:* 541–547 (1983).
52 Kragballe, K.; Voorhees, J.J.: Arachidonic acid in psoriasis. Pathogenic role and pharmacological regulation. Acta derm.-vener., Stockh. suppl. 120, pp. 12–17 (1985).
53 Binder, R.L.; Jonelis, F.J.: Seborrheic dermatitis in neuroleptically induced Parkinson's disease. Archs Derm. *119:* 473–475 (1983).
54 Velen, N.K.; Christiansen, J.V.; Hjorth, N.; Schmidt, H.: Topical metronidazole in the treatment of rosaceae. Cutis, pp. 209–210, Sept. (1986).
55 Boyle, J.: Topical lithium succinate for seborrheic dermatitis. Br. med. J. (Clin. Res.) *292(6512):* 28–30 (1986).
56 Tegner, E.: Seborrheic dermatitis of the face induced by PUVA treatment. Acta derm.-vener., Stockh. *63:* 335–339 (1983).

Sarah Brenner, MD, Department of Dermatology, Ichilov Hospital,
Tel Aviv 64239 (Israel)

Wld Rev. Nutr. Diet., vol. 55, pp. 165–182 (Karger, Basel 1988)

Possible Nutrient Mediators in Psoriasis and Seborrheic Dermatitis

II. Nutrient Mediators: Essential Fatty Acids; Vitamins A, E and D; Vitamins B_1, B_2, B_6, Niacin and Biotin; Vitamin C Selenium; Zinc; Iron

Sarah Brenner, Crystal Horwitz

Department of Dermatology, Ichilov Hospital, Tel Aviv, and
Department of Gastroenterology, Ichilov Hospital, Tel Aviv, Israel

Contents

Introduction

The skin exhibits a limited number of pathognomic lesions irrespective of the cause, be it metabolic or opportunistic. Frank deficiencies of certain nutrients result in skin conditions indistinguishable from those resulting from, say, infection, genetically inherited disorders or 'stress'.

It is seldom, however, that the nutritional status of a patient with a skin disorder but without underlying organic disease is assessed.

The following chapters will examine the role played by various nutrients in maintaining the integrity of the integument, with particular reference to psoriasis and seborrheic dermatitis.

Essential Fatty Acids

Essential fatty acids (EFA) are supplied by our diet, presumably in sufficient amounts under normal circumstances. In cases of malabsorption, malutilization, alcoholism, disease, trauma or imbalanced diets frank deficiencies can occur. Classically, fat-free TPN has resulted in many cases of EFA deficiency [1] although it has been maintained that skin lesions are not manifest if the fat-soluble vitamins are given as supplements [2]. Biochemical changes, however, invariably occur, often within 2 weeks of the fat-free regimen, well before the clinical signs. Plasma levels of linoleic acid are significantly decreased and an abnormal fatty acid, 5,8,11-eicosatrienoic acid makes its appearance [3]. The triene/tetraene ratio (i.e. 5,8,11-eicosatrienoic/arachidonic acid) is elevated and a ratio of 0.2 is said to by symptomatic of an EFA deficiency [4]. In cases of EFA deficiency, oleic acid is utilized as a substrate and is metabolized to this unusual fatty acid [3]. The pathway linoleic acid→arachidonic acid is thus greatly decreased. Additionally, 5,8,11-eicosatrienoic acid is a potent inhibitor of prostaglandin (PG) synthesis [5, 6]. Its concentration is increased in the skin in EFA deficiency [7] and when applied to normal skin it causes a scaly dermatosis with increased DNA synthesis, increased arachidonic acid and decreased PGE_2 levels [8].

In EFA deficiencies a scaly dermatosis, erythema, xerosis and weeping lesions in body folds may be seen. In the experimental animal these conditions can be reversed by topical application of PGE_2 [5] (even though the role of PGE_2 in inflammatory processes has not finally been resolved [9]), linoleic [10] and arachidonic acids [11], or by the addition of linoleic acid to the diet. EFA deficiency causes increased epidermal DNA synthesis and a manifest hyperproliferation [12], for which it is used as an experimental model [13].

The essential fatty acids are also involved in maintaining the skin's barrier functions. Application of linoleic acid to the skin prevents the intradermal water losses that occur in EFA deficiency [14]. Dietary linoleic acid has an essential role in maintaining the epidermal water permeability barrier [15]. When the barrier defenses are weakened the way is opened for opportunistic infections, which have been sometimes identified as the primary cause of psoriasis [16].

The resemblance between seborrheic dermatitis and psoriasis on the one hand and EFA deficiency on the other is quite profound. Not only are the skin lesions similar if not identical, but fatty acid profiles also coincide. In psoriasis plasma arachidonic, linoleic and oleic acids are significantly

Table I. Synthesis of prostaglandins and leukotrienes from fatty acids

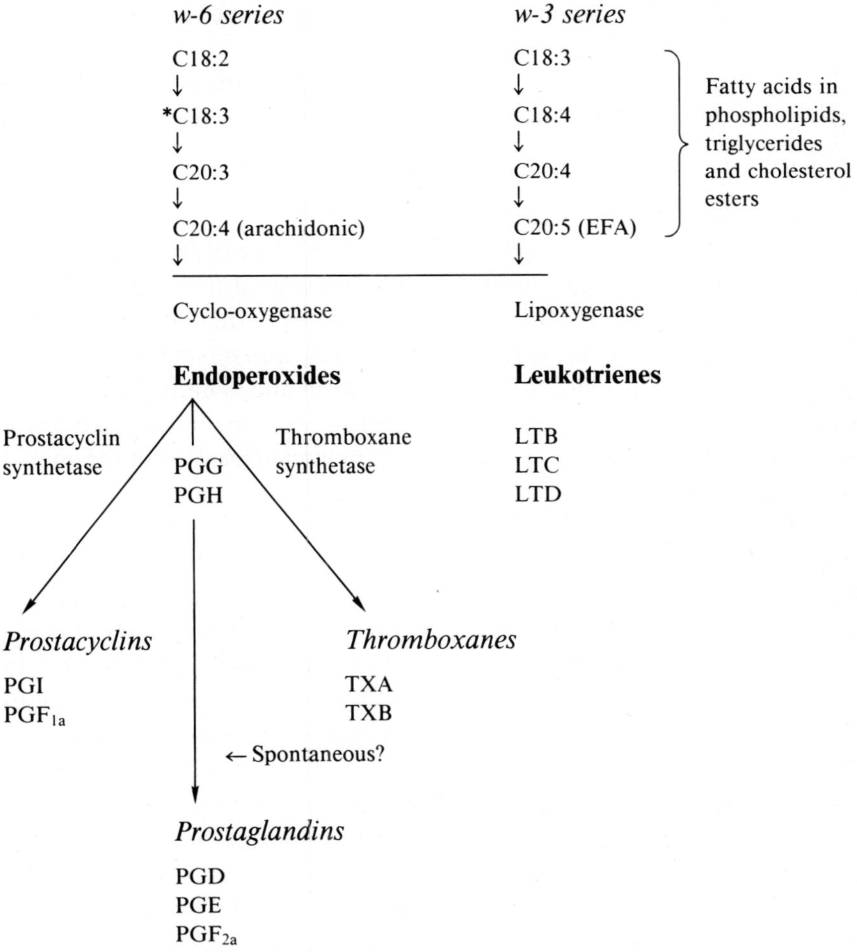

* Evening primrose oil.

reduced [17] while epidermal arachidonic acid is increased [18]. Unfortunately administration of these acids or C18:3n-6, γ-linoleic acid (found in evening primrose oil, see table I) has yielded equivocal results [19]. In any case, the first line of clinical treatment in these disorders is not dietary manipulation or supplementation. Rather it is with anti-inflammatory, antimitotic agents, retinoids and PUVA treatment.

It may be that EFA deficiency of the n-6 series in psoriasis or seborrheic dermatitis is marginal, inconsequential or linked to other nutrient deficiencies that play crucial roles in the metabolism of fatty acids. It is also possible that a marginal deficiency does exist but has gone hitherto undetected.

A comprehensive review of the essential fatty acids has recently been written by Sinclair [20], and new methods of fatty acid status assessed [21].

Vitamin A (Retinol: Vitamin A Alcohol)

Dietary vitamin A is supplied principally by animal fats and diets are supplemented by its addition to various commercial food items such as margarine. Active metabolites can be synthesised in the body from provitamin A (β-carotene mainly) from plant sources. Conversion of β-carotene is almost completely blocked in hypothyroidism and is limited in diabetes and other disorders [22]. High protein diets in experimental animals lower liver stores of vitamin A [23]. These stores in man are sufficient for up to 3 months. Carotene is stored mainly in adipose tissue.

Populations in the developed countries are not regarded as being in need of extra vitamin A and are thought to receive sufficient in their daily diets. Deficiencies occur in disease, malabsorption, malutilization and malnutrition. Vitamin A is transported in the serum by and is found in the skin bound to retinol-binding protein (RBP) [24] which is synthesized in the liver. Free retinol is released from the liver in amounts equal to that of RBP and the complex is bound to prealbumen [25]. Thyroxine also uses this carrier mechanism and thyroxine administration is known to aggravate a vitamin A deficiency [22]. Although the best-known role for vitamin A is in the regulation of vision, it plays an important part in epithelial proliferation and keratinization [26, 27]. Structural lesions caused by vitamin A deficiency (which are the same as those caused by hypervitaminosis A) include hyperproliferation and hyperkeratosis and xerosis of skin and mucous membranes. Psoriatic skin lesions exhibit features identical to those seen in keratinocytes cultured in a surplus of vitamin A [28]. Interestingly, psoriatic patients have not been found to be either RBP- or vitamin A-deficient [29, 30], but their serum and skin β-carotene levels are significantly lower than those of controls [31].

Synthetic retinoids have been used with great success in the treatment of psoriasis and acne, but are not the preferred treatment for the seborrheic dermatitis. Etretinate, but not the carotenoids, increase vitamin A levels in involved psoriatic skin [31]. Etretinate itself resembles the physiologically active *all-trans* retinoic acid (tretinoin:retinoic acid), found in small amounts in animal tissue, but cannot be converted into vitamin A and does not influence its concentration in serum [32]. Its action in psoriasis is unclear. The side effects of the synthetic retinoids in topical and systemic treatment of psoriasis and acne are considerable [33], making their usefulness somewhat limited.

It is of interest to note that tretinoin (retinoic acid) is reported to increase PGE_2 in epithelial tissue [33]. It has been already mentioned that topically applied PGE_2 has beneficial effects of 'psoriatic' lesions. Why such benefits are not bestowed by vitamin A is not known. To confuse the issue further, topical vitamin A can reverse the impairment of wound healing seen in patients on corticosteroids, possibly by restoring normal inflammatory reactions [34].

Ingested vitamin A (and that formed from provitamin A) is found in some cell membranes esterified to fatty acids and is, with vitamin E, important for stabilizing membranes [35]. It is also essential for the synthesis of squalene [36]. The synthetic retinoids act in acne by decreasing the size of the sebaceous glands [33] and the amount of sebum (of which squalene is the principal component) they secrete [37].

Supplements of vitamin A are not effective in the treatment of psoriasis, acne or seborrheic dermatitis, although long-term administration with full control of other dietary factors has not been tried, as far as we know. A review of vitamin A, the retinoids and the skin has recently been published [25].

Vitamin E (α-Tocopherol) and Selenium

Vitamin E has not been implicated in the dermatoses and a deficiency in normal adult populations is not considered prevalent or common. It has been recognized in a number of clinical situations, such as gastric surgery, pancreatic disease, alcoholism, liver disease, abetalipoproteinemia, etc.

An indicator of vitamin E status is the red cell hemolysis test and urine creatine levels, an index of survival times of red blood cells, and is used to detect hemolytic anemia of the newborn. As written above, vitamin E plays a role in stability of cell membranes since the phytyl side chain of the vitamin E molecule interacts with polyunsaturated fatty acids, particularly arachidonic acid, of membrane phospholipids [38]. These side chains inhibit peroxidation of the unsaturated fatty acids [39].

Selenium and vitamin E have synergic actions, both being antioxidants. Selenium deficiency is not acknowledged in Western populations although selenium-poor agricultural soils are recognized as a possible cause for concern. Both vitamin E and selenium are effective oxygen radical traps and thus are counters to lipid peroxidation [40, 41].

It is noteworthy that β-carotene also has antioxidant activity and is a free-radical trap, particularly of highly reactive singlet oxygen [42]. Retinol also has antioxidant properties [43]. Selenium is prescribed (as selenium oxide) in the topical treatment of seborrheic dermatitis. Although the results are not consistent, it is still in the medical armamentarium. Its mode of action is unknown; it might act as an antipromoter of lipid peroxidation of arachidonic acid, via the PG and leukotriene pathways [44].

Vitamin D

Up to the present time vitamin D and its metabolites have not been implicated in skin disorders. Recently there has been some interest manifested in the use of vitamin D_3 in the treatment of psoriasis [45]. The mechanism is unknown.

Some patients benefit from PUVA, whereas others do not. Exposure to sunlight depresses the immune system of the host [46] and this may be an important consideration in the light of suggestions put forward in other sections of this article (see Iron).

Vitamin B_1 (Thiamine) and Niacin

In the past much attention was focused on the neurological and dermatological consequences of vitamin B_1 and niacin deficiencies, principally since the physical signs were so visible. Wet and dry beriberi and pellagra have been known for many years as diseases of malnourished populations, be they peasants in Africa, the East or prisoners of war. Less notice was given to shortfalls of other nutrients, and it is seldom that we can ascribe a deficiency syndrome to the absence of one or two nutrients only.

We cannot yet assign roles to these two vitamins in the skin disorders considered in this article. They may be directly involved, but it is not yet known in what measure. Thiamine pyrophosphate is essential in the synthesis of fatty acids from carbohydrate [47] and collagen repair [48]. Marginal deficiencies in urbanized populations have been described in a disturbingly large number of instances [49–53]. Chronic thiamine deficiency results in prolonged mental and physical deterioration, which takes months of supplementation to rectify [54]. Thus the long-term significance of both vitamin B_1 and niacin deficiency in the etiology of psoriasis and seborrheic dermatitis is unknown. Deprivation in periods of growth and maturation may manifest themselves in later years as failing repair mechanisms.

Biotin

This vitamin is involved in many enzymatic processes, including carboxylations, important for fatty acid and sterol synthesis. It is also involved in transaminations, in the hexokinase system and in purine and pyrimidine synthesis [35].

A biotin deficiency has not been recognized in healthy adults, but infants born to women who are biotin-deficient exhibit severe hair loss, seborrheic dermatitis-like lesions and an exfoliative dermatitis. Identical physiological sequelae can be induced in experimental animals. Biotin has been largely ignored by the medical profession and plays almost no role as yet in the diagnosis and treatment of psoriasis or seborrheic dermatitis.

Vitamins B_2 (Riboflavin) and B_6 (Pyridoxine)

These two vitamins will be considered conjointly, since riboflavin acts as a coenzyme for the synthesis of pyridoxine and because deficiencies of one are frequently accompanied by deficiencies of the other. Dermatoses are known to occur in vitamin B_2 and B_6 deficiencies. Collagen synthesis is also decreased [55]. A pyridoxine deficiency (but not a riboflavin deficiency) causes inflammatory changes in rat skin [56].

Administration of vitamin B_6 antagonists such as desoxypyridoxine, penicillamine and cycloserine cause seborrheic-like skin lesions within 2–3 weeks [57]. Vitamin B_6 is required for the desaturation and elongation of linoleic to arachidonic acid and may influence PG production. There is some evidence for this in the recent finding that asthmatics have low pyridoxal phosphate serum concentrations [58] and that leukotrienes act as triggers for allergic and inflammatory reactions (see above).

In a riboflavin deficiency methemoglobin concentration rises, accelerating the conversion of ferrous to ferric iron [59]. In rats, erythrocytes are more fragile in vitro if they are riboflavin-deficient and these sequelae suggest that riboflavin protects against oxidative damage [60]. Plasma thyroxine levels are reduced in riboflavin deficiency [61, 62] and as hypothyroidism depresses the formation of retinol from β-carotene there exists a complex interdependence between many, if not all, the vitamins. Various studies have shown that marginal deficiencies of vitamins B_2 and B_6 do exist in the USA and the UK [63–66].

Although the nasolabial seborrhea that is a feature of vitamin B_2 and B_6 deficiency seems to indicate that these vitamins should or could play a role in the therapy of psoriasis and seborrheic dermatitis, they do not. In the light of recent reports that vitamin B_6 inhibits the growth of hepatoma cells [67] and melanoma cells [68], it is tempting to speculate that it might yet be used in the treatment of hyperproliferation of certain skin disorders.

Vitamin C (Ascorbic Acid)

Vitamin C has been in the forefront of popular interest for many years and the controversy surrounding its awesome benefits persists. Its functions have been thoroughly documented. Only a few need reemphasis in the context of this article.

Vitamin C is necessary for the synthesis of collagen as it is involved in the hydroxylation of proline [69] and is necessary for the protection of intercellular substances containing collagen [70]. It promotes the immune response [71] and facilitates the absorption of iron [72]. Vitamin C can prevent a selenium-vitamin E deficiency in chicks [72] and increases the bioavailability of selenium [73]. It may increase the conversion of cholesterol to bile acid and thus play a role in cholesterol synthesis [74]. Anemia and hyperkeratotic hair follicles are frequently seen in scurvy. However, seborrheic-like lesions are not a feature in vitamin C deficiency. Several drugs in common use increase vitamin C requirements. These are aspirin, sulfinpyrozone, indomethacin, phenylbutazone, tetracycline, oral contraceptives, corticosteroids and chlorcyclizine [75]. Smoking increases daily requirements [76].

Although it is assumed that we obtain sufficient vitamin C from our diets, according to the US Recommended Dietary Allowances (RDA), there is evidence that certain groups in developed countries, particularly the aged, require supplements [77]. Vitamin C plays no role in clinical therapy for the dermatoses described in this article.

Zinc

Zinc is an essential mineral found in nearly 100 metalloenzymes. The RDA for adults is set at 15 mg/day, identical to that of iron [78]. It is still not certain whether populations in developed countries are subclinically zinc-deficient as intakes have been found to be below the RDA in some instances [79–81].

Zinc is necessary for elongation and desaturation of fatty acids. Arachidonic acid is decreased in the serum of patients with acrodermatitis enteropathica, an inherited disorder in which severe zinc deficiency occurs and which is greatly improved by zinc supplements [82, 83]. In experimental animals an induced zinc deficiency results in decreased desaturation of linoleic and di-homo-γ-linoleic acids, with resultant decreased arachidonic acid synthesis [84] and possibly altered PG formation. In zinc deficiency PGE_2 levels are increased [85]. The lesions of acrodermatitis entheropathica resemble those of psoriasis.

Interrelationships between zinc and vitamin A exist, not yet fully investigated. RBP is zinc-dependent [86] and zinc is necessary for the activity of retinol alcohol dehydrogenase [87]. The retinoids have been used with considerable success in the treatment of psoriasis [88, 89]. Zinc is also implicated in the synthesis of collaged [90] and a deficiency is possibly linked to infantile cutis laxa [91], a disorder of collagen synthesis in the skin.

Deficiencies of zinc are associated with general seborrheic-like dermatoses, involving areas where sebaceous glands are numerous [87, 92], the areas implicated in seborrheic dermatitis. The skin is known to contain high levels of zinc, amounting to about 20% of body stores [93], the epidermis containing 3- to 6-fold higher amounts than the dermis [94].

However, zinc is not in the medical armamentarium in either the topical treatment or systemic treatment of psoriasis and seborrheic dermatitis, even though the dermatological lesions of acrodermatitis enteropathica are so rapidly resolved by zinc supplements, and zinc is beneficial in wound-healing. Nevertheless, zinc may stabilize cell membranes [95] and inhibit lipid peroxidation [96] and thus help preserve its barrier functions. In psoriasis low levels of serum zinc have been reported [96], possibly due to losses by desquamating scale [93], and these low levels may limit absorption of linoleic acid [97]. It should also be borne in mind that intestinal mucosal changes seen in psoriasis [98] may further aggravate nutrient shortfalls.

Iron

The hypotheses presented here reflect the uncertainties and inconsistencies found in the roles played in psoriasis by the PGE series, immune function and nutritional status.

Free radical damage in tissues is catalysed by iron [99] which is released from storage sites within the cells after injury [100]. Metal ions facilitate the decomposition of lipid peroxides into cytotoxic substances [101]. Zinc, however, appears to play a protective role against iron-mediated free radical damage [99].

Soluble iron raises collegenase and PGE_2 concentrations [102] and it is known that PGE_2 concentration is raised in psoriatic skin lesions [103]. Mixtures of iron salts and ascorbic acid (found in vitamin preparations) produce highly damaging hydroxyl radicals [104]. In vivo there are naturally many scavengers and antioxidants which protect against gross damage. The balance between antioxidants and oxidants is important in the etiology of various diseases [105].

Iron depletion decreases tissue injury in a complement- and neutrophil-dependent lung-injury model in the rat [106]. Iron-deficient rats have a very high pain threshold, due in part to the suppression of inflammatory response and pain resulting from the thermal injury inflicted in the experiments [107].

An antioxidant effect in iron depletion has been proposed [108] and inactivation of catalytic iron in inflammatory sites has been suggested as a therapeutic approach [109].

Psoriasis and arthritis are both inflammatory conditions, often associated. Susceptibility to psoriasis implies susceptibility to rheumatic disorders [110]. It is known that iron loading exacerbates rheumatoid arthritis [102, 111]. Dietary manipulation by increasing polyunsaturated fats and decreasing saturated fats ameliorates rheumatoid arthritis [112] either by altering substrates for PG production or by decreasing iron intake. Disordered PG metabolism has been postulated in rheumatoid arthritis [113].

Restrictive diets have been used with some succes in psoriasis [114, 115]. Regression of psoriasis has been noted in inmates in concentration camps suffering from gross malnutrition [Brenner, personal communication]. Is iron implicated in this disorder? We hypothesize that it is. Iron deficiency suppresses the immune response: evidence exists that the enhancement of immunocompetence exacerbates psoriasis [116] and its reduction alleviates it [117]. The success of retinoid treatment in psoriasis may be due to a reduction of local iron stores. Recently a case of vitamin

A-induced anemia has been reported [118]. Psoriasis and seborrheic dermatitis are conditions in which inflammatory processes are uncontrolled. Rheumatoid arthritis responds to iron restriction: is it possible that psoriasis would too? No dietary studies have been done with patients using iron-deficient diets and controlling other important nutrients. It is possible that research along these lines will answer some of the puzzling and exceedingly complex questions that these perplexing skin disorders pose.

Comments

Skin lesions are unsightly but seldom fatal, so the attitude of the treating physician is not influenced by matters of life or death; his task is to alleviate the patient's distress as quickly as possible. No matter that he might have to see his patient repeatedly – so much the better, a cynic might muse. A palliative is readily available so why bother with lengthy and tortuous dietary histories and unacceptable recommendations. What good would they do anyway?

The cardinal question is: Are we what we eat? We are fat because we eat too much; we have headaches and hypertension from accumulated years of high salt intake; we become bumpy and itchy when exposed to too many allergens in our food; we are hyperactive if we consume too many food colorants or coffee; we are nervous, lethargic or insomniac on a vitamin B-deficient diet. Why should we not be scaly, inflamed and desquamated when fatty acid/zinc/iron/vitamin A intake is unbalanced?

Are we being facile in ascribing psoriasis and seborrheic dermatitis to a 'genetic predisposition'? It seems more logical to assume that the conditions are provoked by use or misuse of our daily 'bread'. There is no doubt that the ability to cope with stress (since it is stress that determines the extent of the insult) also plays its role, but surely the integrity of membranes at their molecular level is determined by building and maintenance, namely availability of nutrients? The Chinese have held this 'philosophy' for thousands of years. They believe that food is medicine. Pills and potions are taken only as a last resort – signals of real defeat.

Long-term trials of nutritional manipulation should not be difficult. We might even include nutmeg (it inhibits PGE_2 formation), salicylate-containing foods (for their anti-inflammatory properties) as well as various varieties of fish (for their n-6 and n-3 long chain polyunsaturated acids). It is noteworthy that the Eskimos are virtually free of psoriasis [119]. We await studies in this area with interest.

It may be noteworthy that although the Faroese (who have an incidence of psoriasis comparable to other populations) consume a high amount of marine products, their P/S ratio is not dissimilar to that of the Danes [120], which is 0.24 [121], since they consume large amounts of saturated fats. The P/S ratio of the Eskimos is 0.84 [121].

References

1 Yamanaka, W.K.; Clemans, G.W.; Hutchinson, M.L.: Essential fatty acid deficiency in humans. Prog. Lipid Res. *19:* 187–215 (1981).

2 Maynard, A.T.; Rabert, R.L.: Essential fatty acid deficiency with intravenous hyperalimentation. Fed. Proc. *31:* 717 (1972).

3 Fulco, A.J.; Mead, J.F.: Metabolism of essential fatty acids. VIII. Origin of 5,8,11-eicosatrienoic acid in the fat-deficient rat. J. biol. Chem. *234:* 1411–1416 (1959).

4 Holman, R.T.; Smythe, L.; Johnson, S.: Effect of sex and age on fatty acid composition of human serum lipids. Am. J. clin. Nutr. *32:* 2390–2399 (1979).

5 Ziboh, V.A.; Hsia, S.L.: Effects of prostaglandin E_2 on rat skin. Inhibition of sterol ester biosynthesis and clearing of scaly lesions in essential fatty acid deficiency. J. Lipid Res. *13:* 458–567 (1972).

6 Van Evert, W.C.; Nugteren, D.H.; Van Dorp, D.A.: Inhibition of prostaglandin biosynthesis by 5,8,11-eicosatrienoic acid. Prostaglandins *15:* 267–272 (1978).

7 Ziboh, V.A.; Nguyent, T.T.; McCullough, J.L.; Weinstein, G.D.: Possible role of prostaglandins in scaly dermatosis. Prog. Lipid Res. *20:* 857–865 (1981).

8 Sheretz, E.F.: The skin in essential fatty acid deficiency; in Roe, Nutrition and the skin, chap. 7, pp. 117–130 (Liss, New York 1986).

9 Camp, R.D.R.; Greaves, M.W.: Inflammatory mediators in the skin. Br. med. Bull. (in press, 1987).

10 Prottey, C.; Hartop, P.J.; Press, M.: Correction of the cutaneous manifestations of essential fatty acid deficiency in man by application of sunflower-seed oil to the skin. J. invest. Derm. *64:* 228–234 (1975).

11 Prottey, C.: Essential fatty acids and the skin. Br. J. Derm. *94:* 579–587 (1976).

12 McCullough, J.L.; Schreiber, S.J.; Ziboh, V.A.: Cell proliferation kinetics of epidermis in the essential fatty acid deficient rat. J. invest. Derm. *70:* 318–320 (1978).

13 Lowe, N.J.; Stoughton, R.B.: Essential fatty acid deficient mouse: a model of chronic epidermal hyperproliferation. Br. J. Derm. *96:* 155–162 (1977).

14 Hartop, P.J.; Allenby, C.F.; Prottey, C.: Comparison of barrier function and lipids in psoriasis and essenetial fatty acid-deficient rats. Clin. exp. Dermatol. *3:* 259–267 (1978).

15 Mead, J.F.: Functions of the n-6 and n-3 polyunsaturated fatty acids; in Taylor, Jenkins, Proc. XIII Int. Congr. Nutr., pp. 346–349 (Libbey, London 1986).

16 Ford, G.P.; Farr, P.M.; Ive, F.A.; Shuster, S.: The response of seborrheic dermatitis to ketoconazole. Br. J. Derm. *111:* 603–607 (1984).

17 Vahlquist, C.; Berne, B.; Boberg, M.; Michaelsson, G.; Vessby, B.: The fatty-acid spectrum in plasma and adipose tissue in patients with psoriasis. Archs Derm. Res. *278:* 114–119 (1985).

18 Hammerstrom, S.; Hamberg, M.; Samuelsson, B.; Duell, E.A.; Stawinski, M.; Voorhees, J.J.: Increased concentrations of arachidonic acid, prostaglandins E_2 and F_2 and of 12-*L*-hydroxy-5,8,10,14-eicosatetraenoic acid (HETE) in epidermis of psoriasis: evidence of perturbed regulation of arachidonic acid levels in psoriasis. Proc. natn. Acad. Sci. USA *72:* 5130–5134 (1975).

19 Rasmussen, J.E.: Recent developments in the management of patients with atopic dermatitis. J. Allergy clin. Immunol. *74:* 771–776 (1984).

20 Sinclair, H.M.: Essential fatty acids in perspective. Human Nutr. Clin. Nutr. *38C:* 245–260 (1984).

21 McLean, J.G.; Sinclair, A.J.: Assessment of essential fatty acid status; in Taylor, Jenkins, Proc. XIII Int. Congr. Nutr., pp. 350–352 (Libbey, London 1986).

22 Moore, T.: Vitamin A in the normal individual; in Herriot, Symp. Nutrition, p. 281 (Johns Hopkins University Press, Baltimore 1953).

23 Rechcigl, A.M.; Berger, S.; Loosli, J.K.; Williams, H.H.: Dietary protein and utilization of vitamin A. J. Nutr. *76:* 435–440 (1962).

24 Torma, H.; Vahlquist, A.: Vitamin A uptake by human skin in vitro. Archs Derm. Res. *276:* 390–395 (1984).

25 DiGiovanna, J.J.; Peck, G.L.: Vitamin A and the retinoids; in Roe Nutrition and the skin, pp. 45–62 (Liss, New York 1986).

26 Sauberlich, H.E.; Hodges, R.E.; Wallace, D.L.; Kolder, H.; Canham, J.E.; Hood, J.; Raica, N.; Lowry, L.K.: Vitamin A metabolism and requirements in the human studied with the use of labelled retinol. Vitams Horm. *32:* 251–260 (1974).

27 Wolbach, S.B.; Howe, P.R.: Tissue changes following deprivation of fat-soluble A vitamin. J. exp. Med. *42:* 753 (1925).

28 Fuchs, E.; Green, H.: Regulation of terminal differentiation of cultures human keratinocytes by vitamin A. Cell *25:* 617–625 (1981).

29 Mier, P.D.; Hurk, J. van der: Plasma vitamin A levels in the common dermatoses. Br. J. Derm. *91:* 155–159 (1974).

30 Benoldi, D.; Manfredi, G.; Pezzarossa, E.; Allegra, F.: Retinol binding protein in normal human skin and in cutaneous disorders. Br. J. Derm. *105:* 659–665 (1981).

31 Rollman, O.; Vahlquist, A.: Psoriasis and vitamin A. Archs Derm. Res. *278:* 17–24 (1985).

32 Lauharanta, J.: Vitamin A transport complex during treatment with oral aromatic retinoid (Ro10-9359). Acta derm.-vener., Stockh. *61:* 264–267 (1981).

33 Robertson, D.B.; Maibach, H.J.: Dermatological pharmacology; in Katzung, Basic and clinical pharmacology, chap. 63, pp. 756–772 (Lange Medical Publications, Los Altos 1984).

34 Hunt, T.K.: Effect of vitamin A on reversing the inhibitory effect of cortisone on healing of open wounds in animals and man. Ann. Surg. *170:* 633–638 (1969).

35 Baker, H.; Frank, O.: Clinical vitaminology, chap. XI, pp. 161–168 (Interscience, New York 1968).

36 Roels, O.A.: Present knowledge of vitamin A. Nutr. Res. *24:* 129 (1966).
37 Strauss, J.S.; Stranieri, A.M.: Changes in long-term sebum production from isotretinoin therapy. J. Am. Acad. Derm. *6:* 751–755 (1982).
38 Diplock, A.T.; Lucy, J.A.: The biochemical modes of action of vitamin E and selenium: a hypothesis. FEBS Lett. *29:* 205–210 (1973).
39 Maggio, B.; Diplock, A.T.; Lucy, J.A.: Interaction of tocopherols and ubiquinones with monolayers of phospholipids. Biochem. J. *161:* 111–121 (1977).
40 Bieri, J.G.; Corash, V.; Hubbard, N.: Medical uses of vitamin E. New Engl. J. Med. *308:* 1063–1067 (1983).
41 Flohe, L.: in Pryor, Free radicals in biology, vol. 5, pp. 223–254 (Academic Press, New York 1982).
42 Packer, J.E.; Mahood, J.S.; Mora-Arellano, M.O.; Slater, T.F.; Willson, R.L.; Wolfenden, B.S.: Free radicals and singlet oxygen scavengers: a reaction of β-carotene, diphenyl furan and 1,4-diazobicyclo (2,2,2)-octane. Biochem. biophys. Res. Commun. *98:* 901–906 (1981).
43 Bai, N.J.; Krishnamurthy, S.: Effects of retinol, antioxidants and selenium in tissue catalysis of lipid peroxidation. Ind. J. Biochem. *16:* 249–251 (1979).
44 Ohuchi, K.; Levine, L.: α-Tocopherol inhibits 12-0-tetradecanoyl-phorbol-13-acetate-stimulated deacylation of cellular lipids and prostaglandin production. Biochim. biophys. Acta *619:* 11–19 (1980).
45 Morimoto, S.: A patient with psoriasis cured by α-hydroxyvitamin D_3. Med. J. Osaka Univ. *35:* 51–54 (1985).
46 Parrish, J.A.: Ultraviolet irradiation affects the immune system. Pediatrics, Springfield *71:* 129–134 (1983).
47 White, A.; Handler, P.; Smith, E.L.: in Principles of biochemistry, chap. 20, pp. 542–578 (McGraw-Hill, New York 1973).
38 Alvarez, O.M.; Gilbreath, R.L.: Thiamine influence on collagen during granulation of skin wounds. J. surg. Res. *32:* 24–31 (1982).
49 Harper, C.: Wernicke's encephalopathy: a more common disease than realised. A neuropathological study of 51 cases. J. Neurol. Neurosurg. Psychiat. *42:* 226–231 (1979).
50 Brubacher, G.B.: in Taylor, The importance of vitamins to human health. Vitamin B_1 supply in industrialised countries, chap. 3, pp. 17–26 (MTP Press, Lancaster, 1979).
51 The thiamine statue of the Australian people. Summary of a report to the National Health and Medical Research Council. Med. J. Aust. *i:* 232–235 (1978).
52 Lonsdale, D.; Shamberger, R.J.: Red cell ketolase as an indicator of nutritional deficiency. Am. J. clin. Nutr. *33:* 205–211 (1980).
53 Hatanaka, Y.; Ueda, K.: High incidence of subclinical hypovitaminosis of B_1 among university students found by a field study in Ehime, Japan. Med. J. Osaka Univ. *31:* 83–91 (1981).
54 Williams, R.D.; Mason, H.L.; Power, M.H.; Wilder, R.M.: Induced thiamine (vitamin B_1) deficiency in man. Relation of depletion of thiamine to development of biochemical defect and polyneuropathy. Archs. intern. Med. *71:* 38–53 (1943).
55 Bird, T.A.; Levene, C.I.: Lysyl oxidase: evidence that pyridoxal phosphate is a cofactor. Biochem. biophys. Res. Commun. *108:* 1172–1180 (1982).

56 Bamji, M.S.: Biochemical basis of the skin lesions of vitamins B_2 and B_6 deficiencies; in Roe, Nutrition and the skin, chap. 6, pp. 99–115 (Liss, New York 1986).
57 Rosen, F.; Mihich, E.; Nichol, C.A.: Selective metabolic and chemotherapeutic effects of vitamin B_6 antimetabolites. Vitam. Horm. *22:* 609 (1964).
58 Reynolds, R.D.; Natta, C.L.: Depressed pyridoxal phosphate concentrations in adult asthmatics. Am. J. clin. Nutr. *41:* 684–688 (1985).
59 Hassan, F.M.; Thurnham, D.I.: Effect of riboflavin deficiency on the metabolism of red blood cells. Int. J. Vitam. Nutr. Res. *47:* 349–356 (1977).
60 Thurnham, D.I.; Hassan, F.M.; Powers, H.J.: Effects of riboflavin deficiency in erythrocytes; in Taylor, The importance of vitamins to human health, chap. 4, pp. 27–40 (MTP Press, Lancaster 1979).
61 Nolte, J.; Brdicka, D.; Staudte, H.W.: Effect of riboflavin deficiency on metabolism of the rat in hypothyroid and euthyroid state. Biochim. biophys. Acta *268:* 611–619 (1972).
62 Rivlin, R.S.; Menendez, C.; Langdon, R.G.: Biochemical similarities between hypothyroidism and riboflavin deficiency. Endocrinology *83:* 461–469 (1968).
63 Wyndham, C.T.: Consistency of nutrient consumption patterns in the United States. J. Am. diet. Ass. *78:* 587–595 (1981).
64 Hoorn, R.K.J.; Flikweert, J.P.; Wetserink, D.: Vitamin B_1, B_2 and B_6 deficiencies in geriatric patients, measured by co-enzyme stimulation of enzyme activities. Clinica chim. Acta *61:* 151–162 (1975).
65 Lopez, R.; Schwartz, J.V.; Cooperman, J.M.: Riboflavin deficiency in an adolescent population in New York City. Am. J. clin. Nutr. *33:* 1283–1289 (1980).
66 Buzina, R.; Grgic, Z.; Jusic, M.; Sapunar, J.; Milanovic, N.; Brubacher, C.: Nutritional status and physical working capacity. Human Nutr. Clin. Nutr. *36C:* 429–438 (1982).
67 Disorbo, D.M.; Litwach, G.: Vitamin B_6 kills hepatoma cells in culture. Nutr. Cancer *3:* 216–222 (1982).
68 DiSorbo, D.M.; Wagnette, R.; Nathanson, L.: In vivo and in vitro inhibition of B16 melanoma growth by vitamin B_6. Nutr. Cancer *1:* 43–52 (1985).
69 Barnes, M.J.: The function of ascorbic acid in collagen metabolism. Ann. N.Y. Acad. Sci. *258:* 264–277 (1975).
70 Baker, H.; Frank, O.: Clinical vitaminology; chap. X, pp. 153–160 (Interscience, New York 1968).
71 Prasad, J.S.: Effect of ascorbic acid on plasma iron turnover. Clinica chim. Acta *59:* 101–104 (1975).
72 Siegel, B.V.; Morton, J.I.: Vitamin C and the immune response. Experientia *33:* 393–395 (1977).
73 Combs, G.F., Jr.; Scott, M.L.: Antioxidant effects on selenium and vitamin E function in the chick. J. Nutr. *104:* 1297–1303 (1974).
74 Mutaten, M.; Mykkanen, H.M.: Effect of ascorbic acid supplementation on selenium bioavailability in humans. Human Nutr. Clin. Nutr. *39C:* 221–226 (1985).
75 Ginter, E.: Ascorbic acid in cholesterol and bile acid metabolism. Ann. N.Y. Acad. Sci. *258:* 410–421 (1976).
76 Truswell, A.S.: Vitamins II: ABC of nutrition. Br. med. J. *291:* 1103–1106, 1333–1337 (1985).

77 Schorah, C.J.: Inappropriate vitamin C reserves: their frequency and significance in an urban population, in Taylor, The importance of vitamins to human health, chap. 8, pp. 61–72 (MTP Press, Lancaster 1979).
78 Recommended Dietary Allowances: Natn. Acad. Sci., Washington; 9th rev. ed., pp. 134–136 (1980).
79 Sandstead, H.H.: Zinc nutrition in the United States. Am. J. clin. Nutr. *26:* 1251–1260 (1973).
80 Spring, J.A.; Robertson, J.; Buss, D.H.: Trace elements. 3. Magnesium, copper, zinc, vitamin B_6 and folic acid in the British household food supply. Br. J. Nutr. *41:* 487 (1979).
81 Reilly, C.; Harrison, F.: Zinc, copper, iron and lead in scalp hair of students and non-student adults in Oxford. J. hum. Nutr., Lond. *33:* 248–252 (1979).
82 Cash, R.; Berger, C.: Acrodermatitis enteropathica. Defective metabolism of unsaturated fatty acids. J. Pediat. *74:* 729 (1969).
83 Moynahan, E.J.: Acrodermatitis enteropathica: a lethal inherited zinc deficiency. Lancet *ii:* 399 (1974).
84 Cunnane, C.; Soma, K.; Huang, Y.-S.; Manku, M.S.; Horrobin, D.F.: Micronutrient influences in essential fatty acids. Book of abstracts of the XIIIth Int. Congr. Nutr., p. 106 (Libbey, London 1986).
85 Browning, J.D.; Reeves, P.G.; O'Dell, B.L.: Effect of zinc deficiency and food restriction on the plasma levels of prostaglandin metabolites in male rats. J. Nutr. *133:* 755–759 (1983).
86 Neldner, K.H.: Zinc nutriture and the skin; in Roe, Nutrition and the skin, chap. 8, pp. 131–149 (Liss, New York 1986).
87 Kay, R.G.: Zinc and copper in human nutrition. J. hum. Nutr., Lond. *35:* 25–36 (1981).
88 Kaplan, R.P.; Russell, D.H.; Lowe, N.J.: Etretinate therapy for psoriasis; clinical responses, remission times, epidermal DNA and polyamine responses. Am. Acad. Dermatol. *8:* 95–102 (1983).
89 Honigsman, H.; Wolff, K.: Isotretinoin-PUVA for psoriasis. Lancet *i:* 236 (1983).
90 Fernandez-Madrid, F.; Prasad, A.S.; Oberleas, D.: Effect of zinc deficiency on nucleic acids, collagen, and non-collagenous protein of the connective tissue. J. Lab. clin. Med. *82:* 951–961 (1973).
91 Harpey, J.P.; Jaudon, M.C.; Clavel, J.P.; Galli, A.; Darbois, Y.: Cutis laxa and low serum zinc after antenatal exposure to penicillamine. Lancet *ii:* 858 (1983).
92 Hodges, R.E.: Nutrition in medical practice. Nutrition and the integument, including mucous membranes; in Hodges, Nutrition in medical practice, chap. 12, pp. 265–287 (Saunders, Philadelphia 1980).
93 MacMillen, E.M.; Rowe, D.: Plasma zinc in psoriasis: relation to surface area involvement. Br. J. Derm. *108:* 301–305 (1983).
94 Molokhia, M.M.; Portnoy, B.: Neutron activation analysis of trace elements in the skin. III. Zinc in normal skin. Br. J. Derm. *81:* 759–762 (1969).
95 Chvapil, M.; Ryan, J.N.; Zukowski, C.F.: The effect of zinc on lipid peroxidation in liver microsomes and mitochondria. Proc. Soc. exp. Biol. Med. *141:* 150–153 (1972).
96 Greaves, M.W.; Boyd, T.R.C.: Plasma zinc concentrations in patients with psoriasis, other dermatoses and venous leg ulceration. Lancet *ii:* 1019–1020 (1967).

97 Walldius, G.; Michaelsson, G.; Hardell, L.I.; Aberg, H.: The effects of diet and zinc treatment on the fatty acid composition of serum lipids and adipose tissue and on serum lipoproteins in two adolescent patients with acrodermatitis enteropathica. Am. J. clin. Nutr. *38:* 512–522 (1983).
98 Shuster, S.; Watson, A.I.; Marks, J.: Small intestine in psoriasis. Br. med. J. *iii:* 458–460 (1967).
99 Editorial: Zinc in human nutrition. Lancet *i:* 191–192 (1978).
100 Halliwell, B.; Gutteridge, J.M.C.: Lipid peroxidation, oxygen radicals, cell damage and antioxidant therapy. Lancet *i:* 1396–1397 (1984).
101 McBrien, D.C.H.; Slater, T.F. (eds): Free radicals, lipid peroxidation and cancer (Academic Press, London 1982); quoted in ref. [100].
102 Okazaki, I.; Brinckerhoff, C.E.; Sinclair, J.F.; Sinclair, P.R.; Bonkowsky, H.L.; Harris, E.D.: Iron increases collagenase production by rabbit synovial fibroblasts. J. Lab. clin. Med. *97:* 396–402 (1981).
103 Kozuka, T.; Francis, D.M.; Greaves, M.W.: Arachidonic acid metabolites and the skin. Ann. Acad. Med. *12:* 87–91 (1983).
104 Rowley, D.A.; Halliwell, B.: Formation of hydroxyl radicals from hydrogen peroxide and iron salts by superoxide and ascorbate-dependent mechanisms; relevance in the pathology of rheumatoid disease. Clin. Sci. *64:* 649–653 (1983).
105 Ames, B.N.: Dietary carcinogens and anticarcinogens: oxygen radicals and degenerative diseases. Science *221:* 1256–1264 (1983).
106 Sullivan, J.L.; Till, G.O.; Ward, P.A.: Iron depletion decreases lung injury after systemic complement activation. Fed. Proc. *45:* 452 (1986).
107 Yehuda, S.: Animal behavioural aspects of nutritional iron-deficiency with special reference to learning and cognition (Abstract). 3rd Int. Conf. on Diet and Nutrition, Jerusalem 1986.
108 Doroshow, J.H.; Locker, G.Y.; Myers, C.E.: Enzymatic defenses of the mouse heart against reactive oxygen metabolites: alterations produced by doxorubicin. J. clin. Invest. *65:* 128–135 (1980).
109 Blake, D.R.; Hall, N.D.; Bacon, F.; Dieppe, P.A.; Halliwell, B.; Gutteridge, J.M.C.: Effect of a specific iron-chelating agent in animal models of inflammation. Ann. rheum. Dis. *42:* 89–93 (1983).
110 Brewerton, D.A.: A reappraisal of rheumatic diseases and immunogenetics. Lancet *ii:* 799–801 (1984).
111 Lloyd, K.N.; Williams, P.: Reactions to total dose infusion of iron-dextran in rheumatoid arthritis. Br. med. J. *ii:* 323–325 (1970).
112 Kremer, J.M.; Michalek, A.V.; Lininger, L.; Huyck, D.; et al.: Lancet *i:* 184–187 (1985).
113 Morley, J.: Prostaglandins and lymphokines in arthritis. Prostaglandins *8:* 315–325 (1974).
114 Newborg, B.: Disappearance of psoriatic lesions on the rice diet. N.C. med. J. 253–255 (1986).
115 Hino, A.: Recent clinical experience in a diet regimen based on ecological nutrition and fasting therapy: 16 cases of psoriasis vulgaris. Ann. J. Ass. Ecolo. Nutr. Res. *6:* 7–27 (1982).
116 Quesada, J.R.; Gutterman, J.U.: Psoriasis and alpha-interferon. Lancet *i:* 1466–1468 (1986).

117 Harper, J.I.; Keat, A.C.S.; Staughton, R.C.D.: Cyclosporin for psoriasis. Lancet *ii:* 981–982 (1984).
118 White, J.M.: Vitamin A-induced anaemia. Lancet *ii:* 573 (1984).
119 Kromann, N.; Green, A.: Epidemiological studies in the Upernavik District, Greenland. Acta med. scand. *208:* 401–406 (1980).
120 Olsen, S.F.; Sorensen, T.I.A.; Secher, N.J.; Hansen, H.S.; Jensen, B.; Sommer, S.; Knudsen, L.B.: Intake of marine fat, rich in (n-3)-polyunsaturated fatty acids, may increase birth weight by prolonging gestation. Lancet *ii:* 367–368 (1986).
121 Dyerberg, J.: Fish oil, bleeding time and thrombosis; in Horwitz, Advances in diet and nutrition, pp. 211–217 (Libbey, London 1985).

Sarah Brenner, MD, Department of Dermatology, Ichilov Hospital,
Tel Aviv 64239 (Israel)

Wld Rev. Nutr. Diet., vol. 55, pp. 183–209 (Karger, Basel 1988)

Caries Prevention with Xylitol

A Review of the Scientific Evidence

Albert Bär

Bioresco Ltd., Basel, Switzerland

Contents

Introduction

Dental caries is an infectious and nutrition-related disease which is mediated by bacteria that accumulate in large masses, known as dental plaque, on the teeth in the absence of adequate oral hygiene. Fermentation of common dietary carbohydrates by plaque bacteria leads to the formation of acid end products which accumulate in plaque. The resulting decrease in plaque pH may lead to undersaturation of the plaque with respect to calcium and phosphate ions, to a subsequent loss of mineral from the tooth enamel, and finally to formation of an overt cavity.

Approaches aimed at the prevention or elimination of dental caries include: (1) reduction of acid dissolution of tooth mineral and stimulation of enamel remineralization by fluorides; (2) removal of dental plaque by brushing the teeth, and (3) reduction of the availability of fermentable carbohydrates to the acidogenic dental plaque. While it is generally accepted that the application of fluorides and the regular mechanical removal of dental plaque are appropriate procedures for the prevention of tooth decay, it is not yet fully recognized that the substitution of fermentable sugars by non-acidogenic sugar substitutes makes a significant additional contribution to oral health. Particularly for xylitol, there are by now a large number of published scientific reports available which support the usefulness of this sweetener in dental therapeutics and which emphasize its value for dental health. The scientific data on which this conclusion is based are discussed in the present review.

Rationale for the Use of Non-Fermentable Sweeteners in Caries Prevention

For obvious reasons, fermentable sugars cannot be eliminated completely from our daily food. However, this is not necessary for caries prevention. A number of studies have shown that eating sucrose-sweetened foods with three daily meals results in little or no caries [1]. On the other hand, observations with animals as well as humans indicate that above a certain relatively safe threshold level of sucrose in meals, there exists a rather close relationship between caries formation and the frequency of sucrose consumption between meals [2, 3]. The reason for this association is that critical plaque acidification may occur in response to meals, but that between the meals the cleaning and neutralizing effect of saliva raises the plaque pH to levels where remineralization of the demineralized sites is favoured [4]. However, under conditions of frequent consumption of sugar-containing snacks between the main meals, the plaque pH drops repeatedly and remains for prolonged periods of time below the critical value of 5.7 [5]. As a result, periods of demineralization occur more frequently than recovery phases, and dental caries may subsequently develop. This suggests that substitution of sucrose and other fermentable sugars by non-fermentable sugar substitutes has a considerable caries preventive effect, even if this substitution is limited to snacks and beverages that are consumed between main meals [6–9].

Xylitol is a particularly suitable substance for this purpose because of its non-fermentability, its inhibitory action on the growth and metabolism of certain cariogenic plaque bacteria and its enhancing effect on some natural defence mechanisms against dental caries [10–12].

Fermentability of Xylitol and Other Sugar Substitutes

Sugar alcohols and carbohydrate sweeteners can be divided into three groups according to their fermentability by oral micro-organisms: easily fermented (fructose, glucose, sucrose), slowly fermented (sorbitol, lactitol, maltitol, Lycasin®, polydextrose, palatinose), and non-fermentable (xylitol) [13, 14]. Coupling sugars may be positioned between easily and slowly fermented carbohydrates [15].

Microbiological experiments have clearly shown that xylitol cannot be catabolized into tooth destroying acids by most micro-organisms present in the oral cavity [16–25]. The very few strains that are able to degrade xylitol occur in human dental plaque only in insignificant numbers, and then xylitol degradation proceeds only at an extremely slow rate [26, 27]. Experiments with gnotobiotic rats have confirmed that such bacteria lack significance for caries formation [28, 29]. There have been suspicions that prolonged use of xylitol may result in an increase in xylitol-utilizing bacteria in dental plaque. Experimental results demonstrate, however, that even after 2 years of chronic xylitol consumption, no such adaptive changes occurred [30, 31].

In contrast to xylitol, sorbitol and mannitol are slowly degraded by a number of oral microorganisms, especially by *Streptococcus mutans,* which plays a major role in the development of caries [16, 32–36]. Continuous exposure to sorbitol may even lead to a metabolic adaptation of some plaque bacteria to this substrate and thus to an increased acid production after sorbitol exposure [33, 37, 38]. However, this issue has remained controversial because other investigators have found that the metabolism of sorbitol is rapidly repressible by traces of glucose which, under in vivo conditions, are available to the oral bacteria with nearly every meal [39].

Telemetric recordings of the pH changes in interproximal plaque have also confirmed that under in vivo conditions, administration of xylitol was not associated with any measurable pH drop, whereas exposure to sorbitol or mannitol resulted in some experiments in a slight plaque acidification [23, 40–43].

Effects of Xylitol on Dental Plaque

For the formation of dental plaque, fermentable substrates are required not only as an energy source but also for the bacterial production of structural and adhesive macromolecules like extracellular polysaccharides (glucans and fructans), lipoteichoic acid (LTA) and lectins. These biopolymers play an important role in the accumulation of bacteria on the tooth surface because they determine the adhesion and cohesion of the dental plaque. Several species of oral micro-organisms are able to synthesize such compounds but most experiments on adhesion and polymer formation have been conducted with *S. mutans* because of the high cariogenic activity of this bacterium.

Glucans which are of major importance for the agglutination and adhesion of *S. mutans* to tooth surfaces are produced from sucrose by the action of glucosyltransferases. Glucose and fructose in free form do not serve as substrates for the formation of extracellular insoluble polysaccharides. It is probably for this reason that lower amounts of plaque were found after long-term consumption of fructose [44].

As a rule, *S. mutans* is unable to produce either intra- or extracellular polysaccharides from sugar alcohols [45–47]. However, their presence may modulate the polymer synthesis from sucrose. Thus, xylitol inhibited the in vitro production of insoluble carbohydrates from sucrose by *S. mutans* but increased the production of water-soluble polysaccharides [48, 49]. As a result of these changes in polysaccharide composition, a decrease in the adherence of *S. mutans* to hard surfaces was observed in response to xylitol exposure [50, 51]. However, in a comparative study on different sugar alcohols and various synthetic sucrose analogues, even higher inhibitory effects were found for certain disaccharides, such as isomaltose and xylosylfructoside [52], and similar effects were later reported also for palatinose [53].

Besides carbohydrate polymers, LTA seems to play an important role in plaque formation. This compound consists of a glycolipid covalently linked with a glycerol teichoic acid. LTA is found in cultures of *S. mutans* where it contributes to the adhesiveness of these bacteria because of its negative surface charge. Plaque formed in the presence of sucrose contains relatively high amounts of LTA, but xylitol seems to interfere with the synthesis of this polymer [54, 55]. Sorbitol was reported to have an opposite effect as *S. mutans* grown in a chemostat produced higher amounts of LTA in the presence of fructose and sorbitol than in the presence of glucose

[56]. However, for an interpretation of these results it should be recognized that LTA synthesis is influenced not only by the carbohydrate source, but also by various other factors such as growth rate of the bacteria and pH of the culture medium [57, 58].

The metabolic unavailability of xylitol as an energy source for oral micro-organisms and its possible inhibitory effects on polysaccharide and LTA formation are reflected, under in vivo conditions, by a decreased plaque formation and a reduced plaque adhesion. Depending upon the experimental conditions and the dietary treatment of the corresponding control or comparison groups, these changes were evident to different degrees. Particularly marked effects were observed in studies in which the participating human volunteers were treated with either xylitol or sucrose-sweetened tablets, chewing gum, or mouth rinses [59–64]. In these studies, the application of xylitol was associated with a relative reduction in plaque wet weight of between 32 and 54%. In one of these studies the adhesiveness of the plaque was estimated by determining the plaque-covered areas before and after standardized tooth brushing. In line with the effects on LTA synthesis outlined above the results indicated a lower plaque adhesion in the xylitol exposed subjects [59]. Clinical trials in which the xylitol treatment was compared with a sugar-free placebo treatment, or with other sugar substitutes, such as sorbitol or coupling sugar, produced less dramatic results. For instance, no significant quantitative differences in plaque formation were detected between groups receiving different sugar alcohols or a sugar-free placebo such as unsweetened gum base [65–67]. Nevertheless, plaque weights or plaque scores tended to be lowest after exposure to xylitol [37, 60, 68].

Although dental plaque as a whole is a definite aetiological factor in caries development, it is generally accepted that certain types of plaque bacteria are particularly active in initiating and promoting dental caries. An eminent role in this respect has been ascribed to *S. mutans*. Therefore, any measure that effectively reduces the number of *S. mutans* in the oral cavity would seem to be of caries preventive value. While the frequent ingestion of sucrose gives a competitive edge to *S. mutans* over other plaque bacteria because of its ability to remain metabolically active even in a relatively acidic environment, the regular consumption of xylitol was associated with decreased viable counts of *S. mutans* in dental plaque [31, 69–71]. This advantageous effect of xylitol may not only be a consequence of the non-fermentability of this sugar substitute. Recent studies indicate rather that a toxic metabolite (xylitol-5-phosphate) is formed within the bacterial cells which may interfere with glycolysis and which may also be involved in an energy-consuming futile cycle [72–74]. Somewhat conflicting with this view are, however, the results of three earlier studies in which the consumption of xylitol-sweetened chewing gum was not associated with a significant reduction of *S. mutans* in dental plaque [63, 66, 75].

The ability of bacteria to adapt metabolically to a variety of different substrates is well known. It has been found, for example, that mice and rats fed sugar alcohols for prolonged periods of time, harbour large numbers of bacteria in their intestines which have acquired the capacity to utilize the non-absorbed portion of ingested sugar alcohols as an energy source. On the basis of such observations it has been speculated whether similar adaptational changes could occur in the oral cavity. This would abolish the original advantage of sugar alcohols of being non-fermentable and would make their use questionable in tooth-sparing sweets [76]. For xylitol, this possibility was investigated in several studies with human volunteers consuming xylitol for different periods of time [37, 63, 66, 77]. In none of these investigations was the prolonged use of xylitol associated with an impairment of its non-acidogenic properties. The only change which was noticed was a more frequent occurrence of *S. mutans* strains which were able to tolerate the presence of xylitol but which were still unable to grow at the expense of xylitol [77, 78]. Nevertheless, in terms of total numbers of *S. mutans* a decrease rather than an increase was observed [31, 69, 71]. For xylitol, the hypothesis of an adaptation of the plaque bacteria can, therefore, be dismissed. For sorbitol, however, the corresponding data are less consistent. Data from long-term caries studies with sorbitol in children, for example, have been interpreted as showing such adaptive shifts [76], and results from short-term trials seemed to support this contention even more [37, 67]. In these experiments acid production from sorbitol by plaque was slightly higher at the end than at the beginning of a period with sorbitol exposure of the participants. However, it should be recognized that the bacterial degradation of sorbitol proceeds at a rather slow rate and yields mainly ethanol and formic acid. Lactic acid which has a particularly strong demineralizing effect is formed in comparatively small amounts only [79].

It is unlikely therefore that the observed adaptive shifts increase the cariogenicity of sorbitol to effective levels. Of more concern in this respect are observations of high *S. mutans* counts in subjects with chronic sorbitol consumption because such changes in plaque ecology would generally increase the cariogenicity of dental plaque [69, 80, 81].

Animal Experiments

The relative cariogenicity of various sugar substitutes including xylitol has been investigated in a large number of animal studies under different experimental conditions. In most of these investigations, rats of different strains were used as experimental animals. Typically, the test compounds were given either in admixture to normal or sucrose-containing rodent diets, or they were administered between cariogenic meals using alternative feeding techniques or topical applications. In a few experiments, the sugar substitutes were provided with the drinking water.

Comparative studies on the cariogenicity of sugar alcohols and carbohydrates generally showed an extremely low caries incidence in rats consuming xylitol containing diets (table I) [81–84]. In the same experiments other sugar substitutes, such as sorbitol, mannitol and Palatinit® exhibited a slight cariogenic activity which was, however, clearly below that of sucrose. In xerostomized rats which generally develop rampant caries, xylitol was confirmed to be the least cariogenic sugar substitute [85, 86].

Unfortunately, many though not all of the early rat caries studies on sugar substitutes suffered from some short-comings which made a meaningful interpretation of the results rather difficult. First of all, inappropriate basal diets were often used which, if supplemented with sugar alcohols, produced osmotic diarrhoea because of their high skimmed milk content. As a result, poor health and low weight gains were often observed in the treated animals [87–89]. Second, it was argued that complete substitution of sucrose by sugar substitutes is an unrealistic goal for human nutrition and that, therefore, partial rather than total sugar substitution should be studied in the animal model. In order to overcome these problems, new study protocols with intermittent feeding or topical application of the sugar substitutes were developed [90]. This mode of administration mimics the human eating pattern with three main meals providing all essential nutrients. A variable number of sweet between-meal snacks represents the main cariogenic challenge. Since this procedure results at the same time in lower daily intakes of sugar substitutes, it also reduces or even eliminates the treatment related diarrhoea. In addition, lactose-free basal diets were developed in order to eliminate any interference by this carbohydrate on caries formation and on gastro-intestinal tolerance of the slowly digestible sugar substitutes [91]. Using such improved techniques a number of interesting observations were made (table II). In a study with Osborne-Mendel rats superinfected with *S. mutans* (OMZ 176) and *Actinomyces viscosus*

Table I. Experimental caries studies in rats with continuous administration of xylitol

Exp.	Diet (type)	Inocu-lation[1]	Treatment		Adminis-tration	Days	Reduction of caries[3]	Ref-erence
			control group[2]	xylitol group[2]				
1	natural	not spec.	basal diet	basal diet + 10% Xyl	ad lib. ad lib.	45 90	70[a, b] 80[a, b]	82
2	SSP	SM	50% starch	25% starch + 25% Xyl	18 meals/day	21	37[d]	83
3	SSP	SM (c)	20% Suc 25% Suc	20% Xyl 25% Xyl	18 meals/day 14 meals/day	72 72	95[a, b] 96[a, b]	103
4	2000f	SM, AV	15% Suc 25% Suc	15% Suc + 15% Xyl 25% Suc + 25% Xyl	ad lib. ad lib.	40 40	32[a] 11[a]	87
5	2000	SM (c) SM (d)	50% Suc 50% Suc	50% Suc + 5% Xyl 50% Suc + 5% Xyl	ad lib. ad lib.	42 42	14[a, b], 50[a, c] (6[a, b]), 47[a, c]	91
6	MIT-200	not spec.	67% Suc	57% Suc + 10% Xyl	10 meals/day	42	10[a]	105
7	SSP	SM (c) SM (d)	20% Suc + 5% Glu 20% Suc + 5% Glu	20% Suc + 5% Glu + 5% Xyl 20% Suc + 5% Glu + 5% Xyl	ad lib. ad lib.	66 66	60[a, b] 60[a, b]	91
8	SSP	SM (d) SM (d)	20% Suc + 5% Glu 20% Suc + 5% Glu	20% Suc + 5% Glu + 5% Xyl 20% Suc + 5% Glu + 5% Xyl	ad lib. ad lib.	54 75	55[a, b]; 58[a, c] 40[a, b]; 26[a, c]	98
9	SSP	SM (d)	20% Suc + 5% Glu	20% Suc + 5% Glu + 5% Xyl	ad lib.	62	22[a, b]	99
10	SSP	SM (d)	20% Suc + 5% Glu	20% Suc + 5% Glu + 5% Xyl	12 meals/day	60	51[a, b]; 61[a, c]	104

[1] Rats were inoculated with *Streptococcus mutans* (SM) of serotypes c or d and with *Actinomyces viscosus* (AV).
[2] Abbreviations of test compounds: sucrose (Suc), glucose (Glu), and xylitol (Xyl).
[3] The reduction of caries is expressed in percents of controls. Values in parentheses are increases of caries.
The superscripts signify: [a] T + B + C lesions; [b] mandibular molars; [c] maxillary molars; [d] B lesions.

(Ny 1), the cariogenicity of two topical applications of sucrose, xylitol, sorbose and sorbitol after each of five programmed cariogenic meals per day was investigated. Evaluation of the fissure caries incidence revealed that highest scores were observed in the sucrose and sorbitol treated rats. The number of fissure lesions in the xylitol and sorbose groups did not significantly exceed that of a control group receiving topical applications of plain water [92]. This result demonstrates the non-cariogenicity of topical xylitol applications between meals. Data from a later series of similar experiments allowed the same conclusion [93]. Trials in which xylitol, sorbitol or sucrose were provided with chocolate between low-cariogenic main meals are also in agreement with this result [94, 95].

The question as to whether xylitol can actively suppress dental caries, or whether it is merely not caries-promoting, has been investigated in several animal experiments. One group of studies examined these possibilities by administration of xylitol in admixture to a sucrose-containing cariogenic diet [87, 88, 91, 96–100]. In a typical experiment with addition of xylitol to a cariogenic diet, Osborne-Mendel rats were superinfected with *S. mutans* and received different sucrose-containing diets with or without xylitol for periods of between 42 and 66 days. The diets in which 5% starch was substituted by xylitol produced significantly fewer fissure and smooth surface caries lesions than the same diets without xylitol [91].

In other trials, xylitol was given for different periods of time alternately with, or immediately after, a cariogenic challenge with sucrose-containing meals [88, 90, 93, 99, 101–104]. Overall, the results of these rat experiments suggest that xylitol reduces the cariogenic potential of sucrose if the two sweeteners are administered together [87, 91, 96–100]. Under similar conditions sorbitol and other sugar substitutes were found somewhat less effective in this regard [96, 97, 103], but conflicting results have also been presented [88].

Studies in which xylitol was administered separately between cariogenic meals appear to be particularly relevant as this feeding pattern mimics the human situation most closely. Again, the results suggest some caries-reducing properties of xylitol [99, 101–104]. However, this relatively moderate effect was not detected when highly cariogenic diets were used for the main meals [88, 105]. For the same reason, slight differences in the relative efficiency of the different sugar substitutes were detected in only some cases [103, 104], but not always [102]. Moreover, xylitol exhibited a caries-suppressive effect only if it was fed at the expense of starch. In studies in which xylitol solutions were applied, there were no significant

Table II. Experimental caries studies in rats with intermittent administration of xylitol

Exp.	Inoculation[1]	Main meals[2] all groups	Intermittent meals[2]		Days	Reduction of caries[3]	Reference
			control group	xylitol group			
1	SM (d)	20% Suc + 5% Glu (12 meals/day)	none	33% Xyl (3 meals/day)	60	40[a, b]; 78[a, c]	104
	SM (d)	20% Suc + 5% Glu (12 meals/day)	none	33% Xyl (6 meals/day)	60	67[a, b]; 100[a, c]	104
2	SM (d)	20% Suc + 5% Glu (12 meals/day)	starch (6 meals/day)	33% Xyl (6 meals/day)	70	37[a, b]	104
3	SM (d)	20% Suc + 5% Glu (12 meals/day)	starch (6 meals/day)	8% Xyl (6 meals/day)	70	44[a, b]	104
	SM (d)	20% Suc + 5% Glu (12 meals/day)	starch (6 meals/day)	16% Xyl (6 meals/day)	70	48[a, b]	104
	SM (d)	20% Suc + 5% Glu (12 meals/day)	starch (6 meals/day)	32% Xyl (6 meals/day)	70	47[a, b]	104
4	SM (c)	20% Suc + 10% Glu (4 meals/day)	25% Suc (14 meals/day)	25% Xyl (14 meals/day)	72	96[a, b]	103
5	SM (c)	20% Suc + 10% Glu (16 meals/day)	25% Suc (8 meals/day)	25% Suc + 10% Xyl (8 meals/day)	72	34[a, b]	100

6	SM, AV	20% Suc	water	Xyl (50% soln.)			
		(ad lib.)	(5 × 150 µl topical appl./day)		23	26[a, b]	93
		(ad lib.)	(5 × 200 µl topical appl./day)		23	25[a, b]	
		(ad lib.)	(5 × 200 µl topical appl./day)		23	50[a, b]	
		(ad lib.)	(5 × 150 µl topical appl./day)		23	52[a, b]	
7	SM	56% Suc (3 meals/day)	water (twice after each meal)	Xyl (60% soln.)	35	20	90
8	SM, AV	56% Suc (25 meals/day)	water (twice 150 µl after 5 meals)	Xyl (50% soln.)	20	(30)[a, b]	92
9	SM, AV	65.5% Suc (ad lib.)	starch (ad lib. on alternate days)	starch + 3% Xyl	21	25	101
	SM, AV	65.5% Suc (ad lib.)	starch (ad lib. on alternate days)	starch + 6% Xyl	21	39	101

[1] Rats were inoculated with *Streptococcus mutans* (SM) of serotypes c or d and with *Actinomyces viscosus* (AV).
[2] Abbreviations of test compounds: sucrose (Suc), glucose (Glu), and xylitol (Xyl).
[3] The reduction of caries is expressed in percents of controls. Values in parentheses are increases of caries.
The superscripts signify: [a] T + B + C lesions; [b] mandibular molars; [c] maxillary molars.

Table III. Caries prevention studies with xylitol in humans

Participants (initial age)	Mode of xylitol administration	Duration of study months	Group	Daily xylitol consumption, g	Caries increment Δ DMFS$_{1-4}$	Reference
Adults (27.5 years mean age)	total sugar substitution	24	control group (sucrose)		7.2	111
			comparison group (fructose)		3.8	
			test group (xylitol)	50	0.0	
Adults (22.2 years mean age)	chewing gum	12	control group (sucrose)		2.9	112
			test group (xylitol)	6.25	–1.0	
Children (6–11 years)	confectionary	36	control group (untreated)		7.7	118
			comparison group (fluoride)		6.5	
			test group (xylitol)	14–20	4.2	
Children (6–11 years)	confectionary	32	control group (untreated)		7.2	121
			test group 2 (xylitol)	$\leqslant$20	4.4	
			test group 1 (xylitol)	$\leqslant$20	4.7	
Children (11–12 years)	chewing gum	24	control group (untreated)		2.4	127
			test group 1 (xylitol)	5	2.4	
			test group 2 (xylitol)	7	1.6	
			test group 3 (xylitol)	10	1.0	
Children with high caries susceptibility (11–12 years)	chewing gum	36	control group (untreated)		7.0	127
			test group 1 (xylitol)	5	7.5	
			test group 2 (xylitol)	7	3.5	
			test group 3 (xylitol)	10	1.5	
Children (8–9 years)	chewing gum	12	control group (untreated)		3.35	126
			test group 1 (xylitol)	3.39	1.35	
			test group 2 (xylitol)	0.75	1.76	

differences between the xylitol and the plain water control group [92, 93].

In summary, experimental caries studies in rats indicated that xylitol possesses non-cariogenic properties and that it may even unspecifically promote the remineralization of early caries lesions under certain experimental conditions [106]. The studies also showed that it is not necessary to replace all fermentable carbohydrates in a diet in order to reduce caries. Of course, there is always the question as to whether the results from studies with rodents are relevant for the human situation or not. The actual limitations of the rat model are demonstrated by the fact that caries formation is accelerated in rats by inoculation of the oral cavity with particularly virulent micro-organisms, and by the fact that newly erupted and therefore susceptible teeth are continuously exposed to highly cariogenic diets. All results obtained in the rat model therefore require validation from corresponding adequately controlled studies with human volunteers.

Human Studies

The caries-inducing potential of dietary sucrose and the profound role of the mode of its intake was first observed in the landmark Hopewood House and Vipeholm studies [107, 108]. Many later cross-sectional epidemiological studies have added circumstantial evidence for the cariogenicity of sucrose [for reviews, see 1, 109]. More recently, however, it has been reported that the association between the consumption of sweets and the occurrence of caries has weakened over the past years, perhaps due to the use of fluoride toothpaste and the effect of efficient educational efforts for better oral hygiene [110]. Be that as it may, nutritional measures and in particular restriction of sucrose consumption are still considered valid elements of a caries prevention programme [9]. The potential of this approach was first indicated by the 90% reduction in caries observed after the virtually complete substitution of dietary sucrose with xylitol [111], and by a similar massive reduction in caries achieved through the ingestion of relatively small amounts of xylitol, about 7 g per day in chewing gum, consumed between meals [112]. Similarly, dietary restriction of sucrose was found not only to enhance the effect of other methods but actually to be the most effective means of caries prevention in children [113].

The value of xylitol for caries prevention has been investigated in several longitudinal studies with human volunteers (table III). In the first of the so-called 'Turku Sugar Studies' the effect of an almost complete substitution of dietary sucrose by fructose or xylitol was investigated during a period of 2 years [111]. One hundred and twenty-five subjects with a mean age of about 28 years participated in the study. The subjects were assigned, partly on individual preference basis, to three treatment groups. The sucrose group comprised 35, the fructose group 38, and the xylitol group 52 subjects. Since 10 subjects discontinued or were otherwise excluded from the study, the final results are based on the clinical and radiographic dental examinations of 115 individuals. No significant pretreatment differences were observed between the three treatment groups with regard to age, sex, number of primary and secondary carious surfaces, filled tooth surfaces, extracted teeth, or the DMFS index. After one year of the particular dietary regimens, however, a highly significant reduction of the caries increment (exceeding 85%) was already found for the xylitol group when compared to the sucrose group. After 2 years, the increment in the number of decayed, missing, and filled tooth surfaces was 7.2 in the sucrose group, 3.8 in the fructose group, and 0.0 in the xylitol group. In addition to the increment in terms of the conventional DMFS-index, changes in lesion size and the incidence of secondary caries were also calculated. Irrespective of the mode of calculation, the reduction in the caries incidence exceeded 85% in the xylitol group and 30% in the fructose group compared to the sucrose group.

In a second study, partial substitution of sucrose by xylitol was investigated over a period of one year [112]. The initial study population of 102 subjects, with a mean age of about 22 years, was alloted randomly to the sucrose or the xylitol group. The subjects were instructed to maintain their regular dietary habits and oral hygiene procedures. Instructions were further given to consume between three and seven sticks of chewing gum per day which were sweetened with either sucrose or xylitol, and to keep a complete record of the gum consumption during the study. The final results were based on 100 subjects. At start of the study no relevant differences between the two treatment groups were observed with regard to dental caries and age of the participants. The average consumption of chewing gum was 4.0 sticks per day in the sucrose group and 4.5 in the xylitol group, the difference between the groups not being significant. The frequency of sucrose consumption, calculated before and during the study as the sum of the sucrose-containing products ingested in liquid and solid form, with and between meals, was 4.24 times per subject per day in the sucrose group and 4.94 in the xylitol group. The caries incidence rate was calculated in terms of the conventional DMFS index, as well as in combined quantitative and qualitative terms corresponding to those used in the previous 2-year clinical trial. Regardless of measurement mode, there was at least an 80% reduction in caries increment for the xylitol group compared to the sucrose group [112].

Stimulated by the results of these clinical studies, a number of long-term field trials were subsequently commissioned by the World Health Organization (WHO) as part of its Oral Health Programme. The aim of these trials was to test the caries preventive effectiveness of xylitol under field conditions in different countries and in different nutritional, social and economic environments [114]. By definition the design of these field demonstration programmes differed from that of standard clinical tests in that non-randomized populations with differing baseline caries prevalence were involved. In addition, these trials could not be run under double-blind or even under blind conditions. So far, four such trials have been conducted, one in Thailand [114–116], one in Hungary [116–119] and two in French Polynesia [114, 120, 121].

The trial in Thailand and the first of the two Polynesian trials compared the effectiveness of various chewing gum formulations with the effectiveness of a fortnightly supervised sodium fluoride rinse. One gum formulation contained fluoride and was sweetened with either sucrose or with xylitol and sorbitol. Unfortunately, serious logistical problems were faced in both trials. Nevertheless, the important information from these field trials is that a fluoride-containing chewing gum sweetened with xylitol/sorbitol appears to be effective in controlling caries prevalence. On the other hand, the administration of a fluoride-sucrose chewing gum failed to exhibit any notable preventive effect [114, 116].

More satisfactory conditions prevailed in a 32-month field study with xylitol in French Polynesia and in a 3-year caries preventive programme in Hungary. The specific aim of these studies was to assess the possible caries-prophylactic effect of a partial substitution of sucrose by xylitol under field conditions in caries prone school-age children. Children, 6–12 years of age, enrolled in regular local school programmes, participated in both studies. In Hungary the participating institutions were located in or near Budapest whereas in Polynesia the schools of three separate islands (Maupiti, Rangiroa, Bora-Bora) were included in the study. In both countries, the children of the control group were supplied with tooth brushes and fluoridated toothpaste, and received regular instruction in oral hygiene. Otherwise, they kept their normal dietary habits. The children assigned to the xylitol group were enrolled in an identical oral health programme but, in addition, were provided with xylitol-based sweets, such as chewing gum, candies, chocolate, ice lollies and gumdrops at school on a daily basis. The distribution of sweets was limited, however, to the effect of limiting the individual intake of xylitol to approximately 20 g/day. The consumption of xylitol products was partially supervised but the children always had unrestricted access to conventional, sucrose-based sweets and soft drinks at local commercial outlets or at home. In the Hungarian study a third group was included which received a systemic fluoride treatment with either fluoridated milk or with tap water of naturally high fluoride content. The children of the fluoride group were also regularly instructed in oral hygiene.

Dental caries was scored by experienced examiners before the study began, followed by yearly examinations employing standard diagnostic criteria defined by WHO. In Hungary, additional caries-related parameters (plaque, saliva, nutrition) were determined [122–125]. For statistical assessment, caries data were available from 689 Hungarian children and 468 Polynesian children. At the end of the 32-month Polynesian study, the children of the control group showed an overall mean caries increment of 7.1 decayed, missing or filled surfaces (DMFS). In the xylitol group a mean caries increment of 4.5 DMFS was measured. This statistically significant difference corresponds to a 37% reduction in the caries increment. Calculations of the caries increment rate which adjusted for differences in the number of tooth surfaces at risk, yielded an essentially identical result [121]. In the Hungarian study, a similar reduction in caries increment of 45% was observed for the xylitol group. On the other hand, systemic and/or topical fluoride application alone (fluoride group) had relatively little effect [118]. From these observations it was concluded that regular consumption of xylitol-based sweets enhanced the effectiveness of a caries prevention programme which included regular instructions in oral hygiene and topical fluoride treatment.

Recently, two further field trials were conducted, one in Canada (Montreal) and one in Finland (Ylivieska). Both studies provide evidence that the caries preventive effect of xylitol is dose-dependent. The Canadian study was carried out in 8- and 9-year-old children, randomly assigned to two experimental groups consuming chewing gum with either 15 or 65% xylitol three times daily and to a control group without chewing gum. After one year, the mean DMFS increments were 3.4, 1.8 and 1.4 for the control, the 15 and the 65% xylitol group, respectively [126]. In the Finnish study, a dose-dependent pattern was related to the number of xylitol-containing chewing gums consumed per day. For a period of 2 years, different groups of 11- to 12-year-old children were given one, two and three xylitol-sweetened chewing gums per day corresponding to a maximum daily xylitol intake of 10 g. A control group received no gum. All participants used fluoride for basic caries prophylaxis. Assessment of the caries data showed that daily consumption of two or three xylitol gums resulted in a 33 and 58% reduction of the caries increment when compared with the untreated control group. Consumption of only one gum per day had no effect on caries formation [127].

Taken together, the results of these studies demonstrate that the consumption of even relatively small amounts of xylitol can significantly reduce the formation of new dental caries. Most noticeably, these beneficial effects can occur even in populations which are already treated adequately with fluoride, and which follow normal oral hygiene practices. Thus, it is reasonable to postulate that xylitol, together with fluoride and tooth brushing, should form an integral part of every complete oral health programme.

Suggested Mechanisms for the Caries-Inhibitory Effect of Xylitol

In most animal and human caries studies involving xylitol, a remarkable reduction in the number of newly formed caries lesions was observed in response to the xylitol treatment. In some experiments, even a net reduction of the caries incidence was found, which resulted from a higher number of remineralizations of incipient caries lesions in the xylitol treated groups [98–102, 112]. This result promoted the speculation that xylitol is not only a non-cariogenic sweetener but that it may actually possess cariostatic or even anticariogenic properties. Different mechanisms have been proposed as a possible explanation of such an effect.

Considering de- and remineralization of dental enamel as a direct consequence of changes in the acid-base equilibrium in the tooth environment (plaque fluid, saliva), it was argued that xylitol, due to its organoleptic properties, stimulates salivation, thereby increases plaque pH, and thus promotes remineralization [102]. Other more specific effects of xylitol on the composition of saliva have also been proposed, such as an increase in the concentration of basic amino acids and glycine [128], an increase in total salivary protein in parotid and palatine gland secretions [129, 130], higher activities of amylase [129], carbonic anhydrase [131], lactoperoxidase [132–134] and invertase-like enzymes [135], as well as elevated levels of thiocyanate ions [135, 136]. All these changes indicate a xylitol-mediated enhancement of natural defence mechanisms against caries in terms of a better buffering capacity and a higher bacteriostatic activity of saliva. However, in the absence of cross-sectional, epidemiological data correlating differences in these parameters with caries prevalence, the clinical relevance of the observed changes cannot yet be determined.

Effects of xylitol on the ecology and composition of dental plaque have been described in some detail above. For an explanation of the cariostatic properties of xylitol, the inhibitory and toxic effects of xylitol on *S. mutans* appear to be of particular relevance. In view of the frequent use of xylitol-sorbitol mixtures in sugarless sweets, it is particularly interesting that the combination of sorbitol and xylitol caused an even higher inhibition of *S. mutans* growth than xylitol alone [137]. However, there are also data indicating that some bacteria may acquire resistance against this inhibitory effect of xylitol after long-term exposure to this polyol [138, 139].

Only few investigations have dealt with the possibility that xylitol may affect enamel demineralization or remineralization by direct physico-chemical interaction with solid hydroxyapatite or with calcium ions in solution [105, 140–146]. An in vivo study with implanted enamel slabs showed that neither xylitol nor fluoride have measurable effects on the rehardening of surface softened enamel under conditions favouring remineralization [147]. Different results were, however, obtained in a similar study under conditions favouring demineralization. In this experiment, implanted artificial fissures were treated with solutions of xylitol and sucrose, or with plain water only. Mineral losses and lesion depths in plaque-free and plaque-covered enamel were examined after 16 days of treatment. This study showed a significantly smaller mineral loss in plaque-free enamel treated with xylitol in comparison with the water treatment. No significant differences were found in the plaque covered enamel of the artificial fissures [148]. This result is in line with the findings of earlier in vitro studies on the acid dissolution of enamel in the presence of xylitol [105, 140]. The data suggest that xylitol directly inhibits demineralization of enamel, probably by interfering with the transport of dissolved hydroxyapatite from the lesion to the bulk solution by lowering the diffusion coefficients of calcium and phosphate ions.

Conclusions

In the light of the scientific evidence available at present, xylitol may be regarded as the best of all nutritive sugar substitutes with respect to caries prevention. In many experiments with different oral micro-organisms and with dental plaque it has been shown that xylitol is virtually non-acidogenic. In line with these findings, xylitol did not exhibit any caries-promoting properties in numerous animals trials, as well as in a 2-year clinical study with adult human volunteers. Based on this evidence, the fact that xylitol is non-cariogenic is generally accepted. In several human studies a massive reduction in caries increment was seen in association with the consumption of relatively small amounts of xylitol. These findings led to the hypothesis of a cariostatic or even anticariogenic effect of xylitol. Although a consensus on this subject has not yet been reached, more and more evidence is accumulating which suggests that xylitol is not only an inert sweetener but may indeed be effective in reducing dental caries in different ways. Xylitol was found specifically to inhibit the growth and metabolism of certain strains of cariogenic micro-organisms and also to interfere with the demineralization of tooth enamel in other more direct ways.

On this basis, it is concluded that partial sugar substitution by xylitol is a powerful tool in preventing caries and should be considered together with fluoridation and oral hygiene measures as an equally important factor for the maintenance of individual oral health.

References

1 Newbrun, E.: Sugar and dental caries: a review of human studies. Science *217:* 418–423 (1982).
2 König, K.G.: Caries activity induced by frequency-controlled feeding of diets containing sucrose or bread to Osborne-Mendel rats. Archs oral Biol. *14:* 991–993 (1969).
3 Huxley, H.G.: The effect of feeding frequency on rat caries. J. dent. Res. *56:* 976 (1977).
4 Koulourides, T.; Cameron, B.: Enamel remineralization as a factor in the pathogenesis of dental caries. J. oral Pathol. *9:* 255–269 (1980).
5 Imfeld, T.; Mühlemann, H.R.: Cariogenicity and acidogenicity of food, confectionery and beverages. Pharmacol. Ther. Dent. *3:* 53–68 (1978).
6 Hefti, A.: Zuckerersatzstoffe in der Kariesprophylaxe. Schweiz. med. Wschr. *110:* 269–273 (1980).
7 Mandel, I.D.: Dental caries. Am. Sci. *67:* 680–688 (1979).

8 Scheinin, A.: Sugar substitutes in relation to the incidence of clinical and experimental caries. Pharmacol. Ther. Dent. *3:* 95–100 (1978).
9 Loesche, W.J.: The rationale for caries prevention through the use of sugar substitutes. Int. dent. J. *35:* 1–8 (1985).
10 Mäkinen, K.K.; Scheinin, A.: Xylitol and dental caries. Annu. Rev. Nutr. *2:* 133–150 (1982).
11 Mäkinen, K.K.: Biochemical principles of the use of xylitol in medicine and nutrition with special consideration of dental aspects. Experientia *30:* suppl., pp. 1–160 (1978).
12 Jori, A.: The sugar alcohols: a profile. Adv. Pharmacol. Chemother. *20:* 191–218 (1984).
13 Imfeld, T.: Non-acidogenic and non-cariogenic sugar substitutes and sweeteners; in Guggenheim, Cariology 1983, pp. 147–153 (Karger, Basel 1984).
14 Birkhed, D.; Kalfas, S.; Svensäter, G.; Edwardsson, S.: Microbiological aspects of some caloric sugar substitutes. Int. dent. J. *35:* 9–17 (1985).
15 Ikeda, T.: Sugar substitutes: reasons and indications for their use. Int. dent. J. *32:* 33–43 (1982).
16 Hayes, M.L.; Roberts, K.R.: The breakdown of glucose, xylitol and other sugar alcohols by human dental plaque bacteria. Archs oral Biol. *23:* 445–451 (1978).
17 Edwardsson, S.; Birkhed, D.; Mejare, B.: Acid production from Lycasin®, maltitol, sorbitol and xylitol by oral streptococci and lactobacilli. Acta odont. scand. *35:* 257–263 (1977).
18 Soyer, C.; Frank, R.M.: Influence du milieu de culture sur la croissance du *Streptococcus mutans* ATCC 25175 en présence de différents glucides et leurs dérivés. J. Biol. buccale *7:* 295–301 (1979).
19 Platt, D.; Werrin, S.R.: Acid production from alditols by oral streptococci. J. dent. Res. *58:* 1733–1734 (1979).
20 Noguchi, T.; Mühlemann, H.R.: The effects of some carbohydrates on in vitro growth of *Streptococcus mutans* and *Actinomyces viscosus.* Schweiz. Mschr. Zahnheilk. *86:* 1361/45–1370/54 (1976).
21 Birkhed, D.: Automatic titration method for determination of acid production from sugars and sugar alcohols in small samples of dental plaque material. Caries Res. *12:* 128–136 (1978).
22 Bibby, B.G.; Fu, J.: Changes in plaque pH in vitro by sweeteners. J. dent. Res. *64:* 1130–1133 (1985).
23 Edgar, W.M.; Dodds, M.W.J.: The effect of sweeteners on acid production in plaque. Int. dent. J. *35:* 18–22 (1985).
24 Ziesenitz, S.C.; Hofmeister, M.; Hicks, A.; Siebert, G.: Basic studies on acidogenesis by *Streptococcus mutans.* Caries Res. *19:* 175, abstr. (1985).
25 Zuccato, E.; Pecchioni, A.; Andreoletti, M.; Conca, R.; Ferrara, A.; Mussini, E.: In vivo hydrogen production by breakdown of sugars and sugar alcohols by human oral bacterial flora in relation to carbohydrate cariogenic activity. Mondo Odontostomat. *4:* 25–30 (1986).
26 Gehring, F.; Gülzow, H.-J.: Beitrag zum mikrobiellen Xylitabbau. Dt. zahnärztl. Z. *32:* 580–582 (1977).
27 Gallagher, I.H.C.; Fussell, S.J.: Acidogenic fermentation of pentose alcohols by human dental plaque microorganisms. Archs oral Biol. *24:* 673–679 (1979).

28 Karle, E.J.; Gehring, F.: Palatinit und Xylit im gnotobiotischen Rattenversuch. Dt. zahnärztl. Z. *36:* 673–676 (1981).
29 Karle, E.J.; Gehring, F.: Zur Kariogenität xylitabbauender Keime im gnotobiotischen Versuch. Dt. zahnärztl. Z. *31:* 22–25 (1976).
30 Knuuttila, M.L.E.; Mäkinen, K.K.: Effect of xylitol on the growth and metabolism of *Streptococcus mutans.* Caries Res. *9:* 177–189 (1975).
31 Gehring, F.: Mikrobiologische Untersuchungen im Rahmen der 'Turku sugar studies'. Dt. zahnärztl. Z. *32:* suppl. 1, pp. 84–88 (1977).
32 Gülzow, H.-J.: Vergleichende Untersuchungen über den Abbau von Xylit im menschlichen Speichel. Dt. zahnärztl. Z. *29:* 772–775 (1974).
33 Firestone, A.R.; Navia, J.M.: In vivo measurements of sulcal plaque pH after topical applications of sorbitol and sucrose in rats fed sorbitol or sucrose. J. dent. Res. *65:* 1020–1023 (1986).
34 Westergren, G.; Krasse, B.; Birkhed, D.; Edwardsson, S.: Genetic transfer of markers for sorbitol (*D*-glucitol) metabolism in oral streptococci. Archs oral Biol. *26:* 403–407 (1981).
35 Svensäter, G.; Takahashi-Abbe, S.; Abbe, K.; Birkhed, D.; Yamada, T.; Edwardsson, S.: Anaerobic and aerobic metabolism of sorbitol in *Streptococcus sanguis* and *Streptococcus mitior.* J. dent. Res. *64:* 1286–1289 (1985).
36 Yamada, T.; Takahashi-Abbe, S.; Abbe, K.: Effects of oxygen on pyruvate formate-lyase in situ and sugar metabolism of *Streptococcus mutans* and *Streptococcus sanguis.* Infect. Immunity *47:* 129–134 (1985).
37 Topitsoglou, V.; Birkhed, D.; Larsson, L.-A.; Frostell, G.: Effect of chewing gums containing xylitol, sorbitol or a mixture of xylitol and sorbitol on plaque formation, pH changes and acid production in human dental plaque. Caries Res. *17:* 369–378 (1983).
38 Birkhed, D.; Edwardsson, S.; Svensson, B.; Moskovitz, F.; Frostell, G.: Acid production from sorbitol in human dental plaque. Archs oral Biol. *23:* 971–975 (1978).
39 Slee, A.M.; Tanzer, J.M.: The repressible metabolism of sorbitol (*D*-glucitol) by intact cells of the oral plaque-forming bacterium *Streptococcus mutans.* Archs oral Biol. *28:* 839–845 (1983).
40 Hassell, T.M.: pH-Telemetrie der interdentalen Plaque nach Genuss von Zucker und Zuckeraustauschstoffen. Dt. zahnärztl. Z. *26:* 1145–1154 (1971).
41 Mühlemann, H.R.; Boever, J. de: Radiotelemetry of the pH of interdental areas exposed to various carbohydrates; in McHugh, Dental plaque, pp. 179–186 (Livingstone, Edinburgh 1970).
42 Mühlemann, H.R.; Schneider, P.: The effect of sorbose on pH of mixed saliva and interproximal plaque. Helv. odont. Acta *19:* 76–80 (1975).
43 Maiwald, H.-J.; Banoczy, J.; Tietze, W.: Toth, Z.; Vegh, A.: Die Beeinflussung des Plaque-pH durch zuckerhaltigen und zuckerfreien Kaugummi. Zahn Mund Kieferheilk. *70:* 598–604 (1982).
44 Frostell, G.; Blomqvist, T.; Bruner, P.; Dahl, G.M.; Fjellström, A.; Henrikson, C.-O.; Larje, O.; Nord, C.-E.; Nordenvall, K.-J.; Wik, O.: Reduction of caries in pre-school children by sucrose restriction and substitution with invert sugar. The Gustavsberg study. Acta odont. scand. *39:* 333–347 (1981).
45 Karle, E.J.; Gehring, F.: Palatinit® – ein neuer Zuckeraustauschstoff und seine kariesprophylaktische Beurteilung. Dt. zahnärztl. Z. *33:* 189–191 (1978).

46 Havenaar, R.; Huis in't Veld, J.H.J.; Backer Dirks, O.; Stoppelaar, J.D. de: Some bacteriological aspects of sugar substitutes; in Guggenheim, Health and sugar substitutes. Proc. ERGOB Conf., Geneva 1978, pp. 192–198 (Karger, Basel 1979).
47 Ikeda, T.; Okada, H.; Motoda, R.; Doi, Y.: Effect of xylitol on certain metabolisms of *S. mutans.* Nihon Univ. J. oral Sci. *4:* 84–87 (1978).
48 Wood, J.M.: An investigation of the effect of xylitol on the production of polysaccharide from sucrose by *Streptococcus mutans.* Unpublished report on project P506 for Roche Products Ltd. (1976).
49 Söderling, E.; Alaräisänen, L.; Scheinin, A.; Mäkinen, K.K.: Effect of xylitol and sorbitol on the polysaccharide production and adhesive properties of *Streptococcus mutans.* Caries Res. *21:* 109–116 (1987).
50 Verran, J.; Drucker, D.B.: Effects of two potential sucrose-substitute sweetening agents on deposition of an oral Streptococcus on glass in the presence of sucrose. Archs oral Biol. *27:* 693–695 (1982).
51 Pimbley, D.W.; Whitten, J.; Wood, J.M.: The effect of xylitol on extracellular polysaccharide production, adherence to glass and impedimetric response of *Streptococcus mutans.* Unpublished report on Project P76 for Roche Products Ltd. (1979).
52 Imai, S.; Takeuchi, K.; Shibata, K.; Yoshikawa, S.; Kitahata, S.; Okada, S.; Araya, S.; Nisizawa, T.: Screening of sugars inhibitory against sucrose-dependent synthesis and adherence of insoluble glucan and acid production by *Streptococcus mutans.* J. dent. Res. *63:* 1293–1297 (1984).
53 Takazoe, I.: New trends on sweeteners in Japan. Int. dent. J. *35:* 58–65 (1985).
54 Rölla, G.; Oppermann, R.V.; Bowen, W.H.; Ciardi, J.E.; Knox, K.W.: High amounts of lipoteichoic acids in sucrose-induced dental plaque in vivo. Caries Res. *14:* 235–238 (1980).
55 Assev, S.; Vegarud, G.; Rölla, G.; Addition of xylitol to the growth medium of *Streptococcus mutans* OMZ 176. Effect on the synthesis of extractable glycerolphosphate polymers. Acta pathol. microbiol. scand., B, Microbiol. Immunol. *93:* 145–149 (1985).
56 Hardy, L.; Jacques, N.A.; Forester, H.; Campbell, L.K.; Knox, K.W.; Wicken, A.J.: Effect of fructose and other carbohydrates on the surface properties, lipoteichoic acid production and extracellular proteins of *Streptococcus mutans* Ingbritt grown in continuous culture. Infect. Immunity *31:* 78–87 (1981).
57 Jacques, N.A.; Hardy, L.; Campbell, L.K.; Knox, K.W.; Evans, J.D.; Wicken, A.J.: Effect of carbohydrate source and growth conditions on the production of lipoteichoic acid by *Streptococcus mutans* Ingbritt. Infect. Immunity *26:* 1079–1087 (1979).
58 Knox, K.W.; Wicken, A.J.: Effect of growth conditions on lipoteichoic acid production; in Shockman, Wicken, Chemistry and biological activities of bacterial surface amphyphiles, pp. 229–237 (Academic Press, New York 1981).
59 Rekola, M.: Comparative effects of xylitol- and sucrose-sweetened chew tablets and chewing gums on plaque quantity. Scand. J. dent. Res. *89:* 393–399 (1981).
60 Rateitschak-Plüss, E.M.; Guggenheim, B.: Effects of a carbohydrate-free diet and sugar substitutes on dental plaque accumulation. J. clin. Periodont. *9:* 239–251 (1982).
61 Grenby, T.H.; Bashaarat, A.H.; Gey, K.F.: A clinical trial to compare the effects of xylitol and sucrose chewing gums on dental plaque growth. Br. dent. J. *152:* 339–343 (1982).

62 Mouton, C.; Scheinin, A.; Mäkinen, K.K.: Effect on plaque of a xylitol-containing chewing gum. A pilot study. Acta odont. scand. *33:* 27–31 (1975).

63 Larmas, M.; Scheinin, A.; Gehring, F.; Mäkinen, K.K.: Turku sugar studies. XX. Microbial findings and plaque index values in relation to 1-year use of xylitol chewing gum. Acta odont. Scand. *33:* suppl. 70, pp. 321–336 (1975).

64 Pakkala, U.; Liesmaa, H.; Mäkinen, K.K.: The use of xylitol in the control of oral hygiene in mentally retarded children. Proc. Finn. dent. Soc. *77:* 271–277 (1981).

65 Noguchi, T.; Kitamura, S.; Izumizawa, K.; Fukuda, M.; Kinoshita, S.: Clinical and microbial estimation of the effect of coupling sugar, sorbose and xylitol rinses on the early dental plaque formation in man. Kokubyo-Gakkai-Zasshi *51:* 536–543 (1984).

66 Birkhed, D.; Edwardsson, S.; Ahlden, M.-L.; Frostell, G.: Effects of 3 months frequent consumption of hydrogenated starch hydrolysate (Lycasin®), maltitol, sorbitol and xylitol on human dental plaque. I. Biochemical investigations. Acta odont. scand. *37:* 103–115 (1979).

67 Harjola, U.; Liesmaa, H.: Effects of polyol and sucrose candies on plaque, gingivitis and Lactobacillus index scores. Observations on Helsinki school children. Acta odont. scand. *36:* 237–242 (1978).

68 Birkhed, D.; Edwardsson, S.; Wikesjö, U.; Ahlden, M.-L.; Ainamo, J.: Effect of 4 days consumption of chewing gum containing sorbitol or a mixture of sorbitol and xylitol on dental plaque and saliva. Caries Res. *17:* 76–88 (1983).

69 Loesche, W.J.: The effect of sugar alcohols on plaque and saliva level of *Streptococcus mutans.* Swed. dent. J. *8:* 125–135 (1984).

70 Banoczy, J.; Orsos, M.; Pienihäkkinen, K.; Scheinin, A.: Collaborative WHO xylitol field studies in Hungary. IV. Saliva levels of *Streptococcus mutans.* Acta odont. scand. *43:* 367–370 (1985).

71 Banoczy, J.; Bär, A.; Orsos, M.; Gabris, K.; Szöke, J.; Zimmermann, P.: Salivary levels of *Streptococcus mutans,* lactobacilli and yeasts in children consuming xylitol between meals. Caries Res. (in press).

72 Assev, S.; Rölla, G.: Evidence for presence of a xylitol phosphotransferase system in *Streptococcus mutans* OMZ 176. Acta pathol. microbiol. scand., B, Microbiol. Immunol. *92:* 89–92 (1984).

73 Trahan, L.; Bareil, M.; Gauthier, L.; Vadeboncoeur, C.: Transport and phosphorylation of xylitol by a fructose phosphotransferase system in *Streptococcus mutans.* Caries Res. *19:* 53–63 (1985).

74 Hausman, S.Z.; Thompson, J.; London, J.: Futile xylitol cycle in *Lactobacillus casei.* J. Bact. *160:* 211–215 (1984).

75 Hoerman, K.C.: Enumeration of *Streptococcus mutans* in human dental plaque after chewing gum sweetened with xylitol and sucrose. Pharmacol. Ther. Dent. *4:* 11–19 (1979).

76 Linke, H.A.B.: Sugar alcohols and dental health. Wld Rev. Nutr. Diet., vol. 47, pp. 134–162 (Karger, Basel 1986).

77 Mäkinen, K.K.; Söderling, E.; Hämäläinen, M.; Antonen, P.: Effect of long-term use of xylitol on dental plaque. Proc. Finn. dent. Soc. *81:* 28–35 (1985).

78 Mouton, C.; Dextraze, L.; Mäkinen, K.K.; Greb, U.: Xylitol-resistant isolates of *S. mutans* in xylitol consumers. J. dent. Res. *63:* 212, abstr. 376 (1984).

79 Dallmeier, L.; Bestmann, H.-J.; Kröncke, A.: Über den Abbau von Glukose und Sorbit durch Plaques-Streptokokken. Dt. zahnärztl. Z. *25:* 887–898 (1970).

80 Silver, J.G.; Krasse, B.: Treatment of dental caries monitored by microbial methods: report of two cases. J. Can. dent. Ass. *3:* 211–215 (1985).
81 Karyda-Maniatopoulos, A.-M.: A study of the influence of dietary xylitol on dental caries and *S. mutans* colonization in the rat; thesis, Toronto (1983).
82 Shyu, K.-W.; Hsu, M.-Y.: The cariogenicity of xylitol, mannitol, sorbitol and sucrose. Proc. natn. Sci. Counc. ROC. *4:* 21–26 (1980).
83 Van der Hoeven, J.S.: Cariogenicity of lactitol in program-fed rats (Short communication). Caries Res. *20:* 441–443 (1986).
84 Karle, E.J.; Gehring, F.: Wirkung der Zuckeraustauschstoffe Fruktose, Sorbit und Xylit auf Kariesbefall und Plaqueflora der Ratte. Dt. zahnärztl. Z. *30:* 356–363 (1975).
85 Karle, E.J.; Gehring, F.: Kariogenitätsuntersuchungen von Zuckeraustauschstoffen an xerostomierten Ratten. Dt. zahnärztl. Z. *34:* 551–554 (1979).
86 Karle, E.J.; Büttner, W.: Kariesbefall im Tierversuch nach Verabreichung von Sorbit, Xylit, Lycasin® und Calciumsaccharosephosphat. Dt. zahnärztl. Z. *26:* 1097–1108 (1971).
87 Mühlemann, H.R.; Schmid, R.; Noguchi, T.; Imfeld, T.; Hirsch, R.S.: Some dental effects of xylitol under laboratory and in vivo conditions. Caries Res. *11:* 263–276 (1977).
88 Grenby, T.H.; Colley, J.: Dental effects of xylitol compared with other carbohydrates and polyols in the diet of laboratory rats. Archs oral Biol. *28:* 745–758 (1983).
89 Mühlemann, H.R.; Regolati, B.; Marthaler, T.M.: The effect on rat fissure caries of xylitol and sorbitol. Helv. odont. Acta *14:* 48–50 (1970).
90 Gey, F.; Kinkel, H.T.: Topical application of xylitol and sucrose substitutes after restricted cariogenic meals. J. dent. Res. *57:* 111, abstr. (1978).
91 Havenaar, R.; Huis in't Veld, J.H.J.; Stoppelaar, J.D. de; Backer Dirks, O.: A purified cariogenic diet for rats to test sugar substitutes with special emphasis on general health. Caries Res. *17:* 340–352 (1983).
92 Hefti, A.: Cariogenicity of topically applied sugar substitutes in rats under restricted feeding conditions. Caries Res. *14:* 136–140 (1980).
93 Firestone, A.R.; Schmid, R.; Mühlemann, H.R.: The effects of topical applications of sugar substitutes on the incidence of caries and bacterial agglomerate formation in rats. Caries Res. *14:* 324–332 (1980).
94 Karle, E.J.: Die Kariogenität von Xylit im Tierversuch. Dt. zahnärztl. Z. *32:* S89–S95 (1977).
95 Gehring, F.; Karle, E.J.: Der Saccharoseaustauschstoff Palatinit unter besonderer Berücksichtigung mikrobiologischer und kariesprophylaktischer Aspekte. Z. Ernährungswiss. *20:* 96–106 (1981).
96 Grundberg, E.; Beskid, G.; Brin, M.: Xylitol and dental caries. Efficacy of xylitol in reducing dental caries in rats. Int. J. Vitam. Nutr. Res. *43:* 227–232 (1973).
97 Gehring, F.; Karle, E.J.: Über die kariogenen Eigenschaften von Saccharose und Lactose mit Austauschstoffen. Dt. zahnärztl. Z. *36:* 525–528 (1981).
98 Havenaar, R.; Huis in't Veld, J.H.J.; Stoppelaar, J.D. de; Backer Dirks, O.: Anticariogenic and remineralizing properties of xylitol in combination with sucrose in rats inoculated with *Streptococcus mutans.* Caries Res. *18:* 269–277 (1984).

99 Havenaar, R.: The anti-cariogenic potential of xylitol in comparison with sodium fluoride in rat caries experiments. J. dent. Res. *63:* 120–123 (1984).
100 Havenaar, R.: Potential (anti-) cariogenicity of Lycasin® 80/55 in combination with xylitol and *L*-sorbose in rats; in Havenaar, Sugar substitutes and dental caries, pp. 99–102 (Elinkwijk, Utrecht 1984).
101 Leach, S.A.; Green, R.M.: Effect of xylitol-supplemented diets on the progression and regression of fissure caries in the albino rat. Caries Res. *14:* 16–23 (1980).
102 Leach, S.A.; Green, R.M.: Reversal of fissure caries in the albino rat by sweetening agents. Caries Res. *15:* 508–511 (1981).
103 Havenaar, R.; Drost, J.S.; Stoppelaar, J.D. de; Huis in't Veld, J.H.J.; Backer Dirks, O.: Potential cariogenicity of Lycasin® 80/55 in comparison to starch, sucrose, xylitol, sorbitol and *L*-sorbose in rats. Caries Res. *18:* 375–384 (1984).
104 Havenaar, R.: Effects of intermittent feeding of sugar and sugar substitutes on experimental caries and the colonization of *Streptococcus mutans* in rats; in Havenaar, Sugar substitutes and dental caries, pp. 73–79 (Elinkwijk, Utrecht 1984).
105 Mundorff, S.A.; Bibby, B.G.: Xylitol-sucrose effects on enamel dissolution and rat caries. J. dent. Res. *56:* B136, abstr. (1977).
106 Leach, S.A.; Agalamanyi, E.A.; Green, R.M.: Remineralisation of the teeth by dietary means; in Leach, Edgar, Demineralisation and remineralisation of the teeth, pp. 51–73 (IRL Press, Oxford 1983).
107 Harris, R.: Biology of the children of the Hopewood House, Bowral, Australia. Observations of dental caries experience extending over five years (1957–1961). J. dent. Res. *42:* 1387–1399 (1963).
108 Gustafsson, B.E.; Quensel, C.-E.; Swenander Lanke, L.; Lundqvist, C.; Grahnen, H.; Bonow, B.E.; Krasse, B.: The Vipeholm dental caries study. The effect of different levels of carbohydrate intake on caries activity in 436 individuals observed for five years. Acta odont. scand. *11:* 232–263 (1953).
109 Rugg-Gunn, A.J.; Edgar, W.M.: Sugar and dental caries: a review of the evidence. Community dent. Hlth *1:* 85–92 (1984).
110 Sundin, B.; Birkhed, D.; Granath, L.: Is there not a strong relationship nowadays between caries and consumption of sweets? Swed. dent J. *7:* 103–108 (1983).
111 Scheinin, A.; Mäkinen, K.K.; Ylitalo, K.: Turku sugar studies. V. Final report on the effect of sucrose, fructose and xylitol diets on the caries incidence in man. Acta odont. scand. *33:* suppl. 70, pp. 67–104 (1975).
112 Scheinin, A.; Mäkinen, K.K.; Tammisalo, E.; Rekola, M.: Turku sugar studies. XVIII. Incidence of dental caries in relation to 1-year consumption of xylitol chewing gum. Acta odont. scand. *33:* suppl. 70, pp. 307–316 (1975).
113 McDonald, S.P.; Cowell, C.R.; Sheiham, A.: Methods of preventing dental caries used by dentists for their own children. Br. dent. J. *151:* 118–121 (1981).
114 Barmes, D.; Barnaud, J.; Khambonanda, S.; Sardo Infirri, J.: Field trials of preventive regimes in Thailand and French Polynesia. Int. dent. J. *35:* 66–72 (1985).
115 Khambanonda, S.; Chadravejjsmarn, R.; Barmes, D.E.; Sardo Infirri, J.; Möller, I.: Prévention de la carie dentaire en Thaïlande: trois produits fluorés soumis à des tests comparatifs. J. Biol. buccale *11:* 255–263 (1983).
116 Scheinin, A.: Field studies on sugar substitutes. Int. dent. J. *35:* 195–200 (1985).
117 Scheinin, A.; Banoczy, J.: Collaborative WHO xylitol field studies in Hungary. An overview. Acta odont. scand. *43:* 321–324 (1985).

118 Scheinin, A.; Banoczy, J.; Szöke, J.; Esztari, I.; Pienihäkkinen, K.; Scheinin, U.; Tiekso, J.; Zimmermann, P.; Hadas, E.: Collaborative WHO xylitol field studies in Hungary. I. Three-year caries activity in institutionalized children. Acta odont. scand. *43:* 327–347 (1985).
119 Scheinin, A.; Pienihäkkinen, K.; Tiekso, J.; Banoczy, J.; Szöke, J.; Esztari, I.; Zimmermann, P.; Hadas, E.: Collaborative WHO xylitol field studies in Hungary. VII. Two-year caries incidence in 976 institutionalized children. Acta odont. scand. *43:* 381–387 (1985).
120 Hefti, A.F.; Kandelman, D.P.: Efficacy of partial sugar substitution by xylitol in caries prevention. Caries Res. *20:* 184, abstr. (1986).
121 Kandelman, D.; Bär, A.; Hefti, A.: Collaborative WHO xylitol field study in French Polynesia. I. Baseline prevalence and three-year caries increment. Caries Res. *22* (1988).
122 Banoczy, J.; Scheinin, A.; Pados, R.; Ember, G.; Kertesz, P.; Pienihäkkinen, K.: Collaborative WHO xylitol field studies in Hungary. II. General background and control of the dietary regimen. Acta odont. scand. *43:* 349–357 (1985).
123 Pienihäkkinen, K.; Gabris, K.; Nyarasdy, I.; Rigo, O.; Scheinin, A.; Banoczy, J.: Collaborative WHO xylitol field studies in Hungary. III. Longitudinal counts of lactobacilli and yeasts in saliva. Acta odont. scand *43:* 359–365 (1985).
124 Szöke, J.; Pienihäkkinen, K.; Esztari, I.; Banoczy, J.; Scheinin, A.: Collaborative WHO xylitol field studies in Hungary. V. Three-year development of oral hygiene. Acta odont. scand. *43:* 371–376 (1985).
125 Kertesz, P.; Schuder, L.; Szöke, J.; Gintner, Z.; Pienihäkkinen, K.; Scheinin, A.; Banoczy, J.: Collaborative WHO xylitol field studies in Hungary. VI. Changes in the carbohydrate to protein ratio of dental plaque. Acta odont. scand. *43:* 377–380 (1985).
126 Kandelman, D.; Gagnon, G.; Herr, P.: Effect on dental caries of xylitol chewing gums in preventive programs. J. dent. Res. *65:* 199, abstr. (1986).
127 Isokangas, P.; Alanen, P.; Tiekso, J.; Mäkinen, K.K.: A longitudinal study on xylitol chewing gum in caries prevention. Abstr. Annu. Session of AADR, 1987.
128 Mäkinen, K.K.; Lönnberg, P.; Scheinin, A.: Turku sugar studies. XIV. Amino acid analysis of saliva. Acta odont. scand. *33:* suppl. 70, pp. 277–286 (1975).
129 Bird, J.L.; Baum, B.J.; Mäkinen, K.K.; Bowen, W.H.; Longton, R.W.: Xylitol-associated changes in amylase and protein content of monkey parotid saliva. J. Nutr. *107:* 1763–1767 (1977).
130 Mäkinen, K.K.; Virtanen, K.K.; Söderling, E.; Kotiranta, J.: Effect of xylitol-, sucrose-, and water-rinses on the composition of human palatine gland secretions. Scand J. dent. Res. *93:* 253–261 (1985).
131 Mäkinen, K.K.; Virtanen, K.K.; Mäkinen, P.-L.; Söderling, E.; Neva, M.; Sinisalo, T.; Kölling, D.: Effect of single doses of sucrose and xylitol on human parotid saliva. Proc. Finn. dent. Soc. *80:* 186–200 (1984).
132 Mäkinen, K.K.; Virtanen, K.K.; Mäkinen, P.-L.; Neva, M.; Söderling, E.; Sinisalo, T.; Kölling, D.: Enzyme levels of parotid saliva after stimulation with xylitol and sucrose. Caries Res. *14:* 164–165, abstr. (1980).
133 Mäkinen, K.K.; Tenovuo, J.; Scheinin, A.: Xylitol-induced elevation of lactoperoxidase activity in human saliva. J. dent. Res. *55:* 652–660 (1976).
134 Mäkinen, K.K.; Söderling, E.; Kölling, D.: Secretory activity of salivary gland slices in the presence of polyols. Proc. Finn. dent. Soc. *79:* 218–223 (1983).

135 Mäkinen, K.K.; Kölling, D.; Mäkinen, P.-L.: The effect of high oral dosage of xylitol and sucrose on the biochemical properties of whole saliva in human volunteers after long-term regular consumption of xylitol. Proc. Finn. dent. Soc. *77:* 262–270 (1981).
136 Mäkinen, K.K.; Scheinin, A.: Turku sugar studies. VII. Principal biochemical findings on whole saliva and plaque. Acta odont. scand. *33:* suppl. 70, pp. 129–171 (1975).
137 Assev, S.; Rölla, G.: Sorbitol increases the growth inhibition of xylitol on *Strep. mutans* OMZ 176. Acta pathol. microbiol. scand., B, Microbiol. Immunol. *94:* 231–237 (1986).
138 Gauthier, L.; Vadeboncoeur, C.; Mayrand, D.: Loss of sensitivity to xylitol by *Streptococcus mutans* LG-1. Caries Res. *18:* 289–295 (1984).
139 Assev, S.; Scheie, A.A.: Xylitol metabolism in xylitol-sensitive and xylitol-resistant strains of streptococci. Acta pathol. microbiol. scand., B, Microbiol. Immunol. *94:* 239–243 (1986).
140 Arends, J.; Christoffersen, J.; Schuthof, J.; Smits, M.T.: Influence of xylitol on demineralization of enamel. Caries Res. *18:* 296–301 (1984).
141 Vissink, A.; 's-Gravenmade, E.J.; Gelhard, T.B.F.M.; Panders, A.K.; Franken, M.H.: Rehardening properties of mucin or CMC containing saliva substitutes on softened human enamel. Caries Res. *19:* 212–218 (1985).
142 Mäkinen, K.K.; Söderling, E.: Solubility of calcium salts, enamel and hydroxyapatite in aqueous solutions of simple carbohydrates. Calcif Tissue Int. *36:* 64–71 (1984).
143 Scheinin, A.; Scheinin, U.; Glass, R.L.; Kallio, M.-L.; Söderling, E.: Xylitol-induced changes of enamel hardness in the human mouth. J. dent. Res. *60:* suppl. A, pp. 514, abstr. (1981).
144 Koulourides, T.; Bodden, R.; Keller, S.; Manson-Hing, L.; Lastra, J.; Housch, T.: Cariogenicity of nine sugars tested with an intraoral device in man. Caries Res. *10:* 427–441 (1976).
145 Essig, M.E.; Bodden, W.R.; Bradley, E.L.; Koulourides, T.; Housch, T.: Enamel microhardness change and plaque pH measurements in an intra-oral model in humans. J. dent. Res. *64:* 1065–1068 (1985).
146 Rekola, M.: Approximal caries development during 2-year total substitution of dietary sucrose with xylitol. Caries Res. *21:* 87–94 (1987).
147 Smits, M.T.; Arends, J.: Influence of xylitol and/or fluoride-containing toothpastes on the remineralization of surface softened enamel defects in vivo. Caries Res. *19:* 528–535 (1985).
148 Smits, M.T.: Xylitol and dental caries; thesis, Groningen (1987).

Dr. Albert Bär, Bioresco Ltd., Clarastrasse 12, CH–4058 Basel (Switzerland)

Wld Rev. Nutr. Diet., vol. 55, pp. 210–245 (Karger, Basel 1988)

Nitrogen Utilisation by Ruminant Animals

Khaled El-Shazly

Faculty of Agriculture, University of Alexandria, Egypt

Contents

I. Introduction

Ruminant animals, e.g. cattle, buffalo (*Bos bubalis*), sheep, goats and pseudo-ruminants, e.g. camels, llama, hippopotamus and marsupials are characterised by a complex stomach, part of which, the reticulo-rumen (in true ruminants), is inhabited by a large number of micro-organisms; bacteria, protozoa and fungi. Feed constituents, i.e. carbohydrates, proteins and lipids, are subjected primarily to the rumen microbial activities where they are fermented, providing the necessary energy (rich energy phosphate bonds, e.g. adenosine triphosphate; ATP) for the rumen micro-organisms to perform their functions and propagate. The products of fermentation, i.e. volatile fatty acids including mainly acetic, propionic and butyric acids but also in much smaller concentrations, formic acid, C_5, C_6 and isomers (branched chain volatile fatty acids) and microbial proteins which are transferred to the other parts of the alimentary tract, are utilized by the host animal.

Ruminants are characterized by their activities to digest cellulose through microbial cellulases and to synthesise proteins from non-protein nitrogen (NPN). At the same time they are able to synthesize a number of vitamins, especially group B complex vitamins.

Recent studies have been directed towards improving microbial protein synthesis so as to increase the utilization of available amino acids in relation to VFA produced in the rumen.

II. Utilisation of Non-Protein Nitrogen and Proteins by Ruminants

Recent studies have led to the suggestion of the following system for best utilization of nitrogenous compounds by ruminants.

(1) Maximization of NPN, e.g. urea, biuret, in ruminant diets ascertaining maximum possible utilization of microbial protein synthesis, reducing urea losses in the urine.

(2) Minimizing dietary protein intake of low degradability in the rumen (non-soluble, bypass protein) to provide the essential amino acids complimentary to those of microbial origin.

A. Non-Protein Nitrogen

Since World War 1, the necessity of replacing expensive protein-N by NPN has been realized by Germany who manufactured urea and developed the use of NPN compounds in ruminant nutrition [143]. Rumen micro-organisms have ureolytic activities resulting in ammonia and carbon dioxide. Ammonia is used primarily by bacteria for the synthesis of microbial protein. Protozoa and fungi are probably less important in that respect [20]. The rate of urea hydrolysis by rumen bacteria can be as high as 1 g/l/h [89]. It is slightly less in the rumen. Ammonia reaches a peak after 1 h from urea administration, and other NPN sources, e.g. biuret-soluble proteins, e.g. casein and insoluble proteins, e.g. zein, have slower rates of hydrolysis, respectively [32, 108].

The economic value of using urea in the feeding of ruminants is indicated from its nitrogen content (46%) as compared to protein nitrogen (16%). Urea has 287% equivalent protein as compared to cottonseed meal at 40% (in Egypt, the cost of one unit of cottonseed nitrogen was estimated at 1.75 times that of urea-N equivalent).

However, feeding urea to ruminants has its limitations as well as its risks if overfed. The fast release of NH_3 which could be absorbed into the blood [107] and the lymph [37] to reach the central nervous tissue can cause toxicity and death. Ammonia-N concentrations above 100 mg/dl rumen liquor, 2 mg/dl blood plasma and ruminal fluid of pH 8 are indicative of toxicity. These signs could be reversed by acidification before the onset of tetany [15].

Most scientists agree that 30% of protein-N could be satisfactorily replaced by urea-N with no adverse effect [96]. However, there are other experiments where producing animals were maintained on diets in which urea-N substituted all protein-N [126] for growing animals and for dairy cattle [157] producing an everage of 20 litres/day.

1. Hydrolysis of NPN by Ruminants. Urea is hydrolysed in the rumen by the enzyme urease produced by certain rumen micro-organisms which were separated by Hobson [78] and Cook [45]. Biuret is decomposed by biuretase to ammonia, carbon dioxide and urea which is hydrolysed by urease [155]. Seven strains of *Pseudomonas aeruginosa* were found to hydrolyze biuret.

Biuret and other NPN sources, e.g. triuret, cyanuric acid complexes with formaldehyde or molasses, caused retarded release of ammonia [118].

Urea is only useful to ruminants in that it is hydrolysed into ammonia which is used for microbial protein synthesis. Ammonia is one of the requirements of many rumen bacteria [34]. Amino acids may also be required by other bacteria for efficient growth [9]. Amonia concentration in the rumen is the resultant of production by rumen micro-organisms, synthesis of microbial proteins and absorption through the reticulo-rumen wall. Houpt [82] showed that ammonia is absorbed at a rate of 2.4 mmol/h. Urea-N returning to the rumen was estimated at 1–5 g/day (3.26 g) urea or 54 mmol/day or 2.26 mmol/h. Seven hundred milligrams return via saliva (46.6%), the rest amounting to 53.4% via the rumen wall [141]. It is clear, therefore, that the loss of urea is negligible, especially when nitrogen intake is low. Kennedy and Milligan [91] estimated that 23–92% of plasma urea is recycled to the digestive tract depending on the level of nitrogen intake.

Losses in the urine increase with increased levels of N intake causing a drop in nitrogen economy to the animal [3]. Houpt [83] estimated that a maximum of 6 and 24 g/day urea nitrogen recycled to the rumen in sheep and cattle, respectively. He postulated that the first step in urea transfer occurs through the basal layer of the epithelium where only a small portion of urea passes undergraded to the rumen where it is rapidly hydrolysed by the mircrobial urease into ammonia. The major part of the blood urea is hydrolysed in the epithelial tissue perhaps in the cornified part under the influence of microbial urease which penetrates through the epithelium. Ammonia diffuses into the rumen where the pH is slightly acidic and it is transferred to the ammonium ions which are less absorbable than ammonia. The studies which led to the above theory were carried out using the Pavlov pouch technique [92].

Ahmed et al. [5], measured urea recycling in 2 Barki rams given maintenance requirements at 50% of the ARC-N requirements. Ammonia-N concentration was determined at intervals. Regression equations were formulated:

$$Y = -4.76 + 1.01x, \text{ and}$$
$$Y = 5.75 + 1.13x,$$

where Y = rate of urea infused intraruminally (mg N/min), and x = ruminal ammonia-N concentration measured at the time of intravenous infusion of urea.

x was found to be 6.19, therefore, urea recycling was 2.15 and 3.57 g N/day in the 2 rams, respectively. The values were of the same order as found by Weston and Hogan [163] and Houpt [82]. When sheep were deprived of water, urea transfer from plasma to the rumen were reduced to 0.95 g urea-N/day. During watering recycling was of the order of 2.86 g urea-N per day [27]. Nitrogen retention was improved during water restriction. This is in contradiction with the results of Utley et al. [156] who showed that intravenous infusion of urea in water-deprived animals increased urea recycling. Harmeyer and Martens [72] found that entry rates of urea into plasma (the quantity of urea synthesised or liberated into the body pool per unit time) vary from 3.7 to 82 $\mu mol/min \cdot kg^{0.75}$ depending on dietary conditions. It reflects the ability of ruminants to adapt the nitrogen metabolism to different dietary conditions. Daily N intake is, therefore, the major determinant of entry rate of urea ($r^2 = 0.77$) (fig. 1). No relation was found between plasma urea concentration and entry rate (measured using ^{14}C-labelled urea). On the other hand, a positive

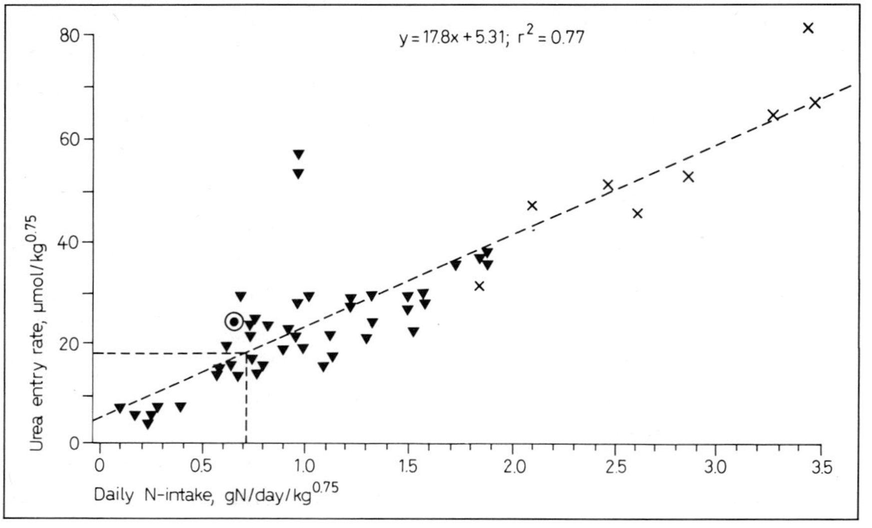

Fig. 1. Relationship between daily N intake and rate of urea entry in goats (●), sheep (▼), and cattle (×). After Harmeyer and Martens [72].

relationship was observed between urea concentration in plasma and urea pool in goats ($r^2 = 0.67$).

Low protein-high energy diets facilitate the return of urea to the gastrointestinal tract which is interpreted as a change of permeability to urea. Potthast et al. [133] indicated that the rate of degraded urea entering the rumen increased with increasing energy level, 27–34% of degraded urea entered the rumen when sheep were fed on 400 g dried grass/day. This increased to 72% when 300 g of saccharose were added to the diet. Similarly, when flaked barley was included in a pelleted-grass diet, there was a significant decrease in urea synthesis rate and a significant increase in recycled urea degraded in the rumen (from 2.3 to 3 g N/day). The permeability of the rumen wall to urea was also significantly increased by giving flaked barley to sheep (121).

B. Degradation of Proteins in the Rumen

Feed proteins mostly of plant origin are subjected to proteolysis by rumen micro-organisms. Proteolytic activities were recognised early by Sym [150] in all fractions of the rumen. Highest activities resided with heavier fractions of rumen bacteria and protozoa. McDonald [180] showed that ammonia reached a peak at 4–5 h, when casein or gelatin were degraded in the rumen. Zein, an insoluble protein, was only slightly degraded and was similar to hay protein in the NH_3 curve. McDonald [109] calculated that 40% of the zein protein are degraded in the rumen, the rest escaped degradation and passed to the duodenum. Casein was almost completely degraded in the rumen (97%) [110].

A certain parallelism was found between solubility in vitro and ammonia concentration in the rumen for a number of protein sources [61]. *Vicea faba* proteins and casein showed highest solubility in vitro and had the highest ammonia concentration in the rumen. Proteins which showed low solubility and degradability in the rumen incurred the greatest nitrogen retention by sheep. Meat meal which showed the lowest solubility in vitro and very low NH_3-N concentration in the rumen gave, unexpectedly, the lowest N retention by sheep. Blackburn and Hobson [25] also studied the relationship between rates of protein digestion by rumen bacteria and solubility in vitro. Hendrickx and Martin [74] illustrated the relationship between solubility of proteins and the extent of their digestion in the rumen. Recent evidence indicates that in vitro solubility alone does not accurately reflect degradability of proteins. Solubility plus in situ measurements together may reflect more accurately the extent of protein degradation [46, 169]. A prediction equation for estimating rumen bypass proteins on the basis of solubility and in situ disappearance rates gave high correlation ($r = 0.984$). Rumen protein degradation rates for casein, soybean, cottonseed, and corn gluten meals were 103, 85, 76 and 54% at 3,000 g/day protein supplements for Holstein steers. At 4,000 g/day, the degradation estimates were lower, 82, 39, 56, 39 and 30% for the same supplements plus meat and bone meal, respectively [168].

Proteins are degraded into amino acids and these are deaminated producing NH_3, volatile fatty acids, e.g. acetic propionic and butyric acid as well as CO_2. The concentration of amino acids in the rumen is quite low relative to ammonia suggesting higher rates of deamination than proteolysis [58]. Amino nitrogen in the rumen varied from 0.05 to 0.2 mmol while ammonia reached 40 mmol/l after 100 g of gelatin was introduced into the rumen [97]. Branched chain fatty acids increased in rumen liquor with increasing protein in the ration [59]. Cellulolytic *Bacteroides* succinogenase from the rumen was found to have requirements for branched isomers of volatile fatty acids, e.g. isobutyric, isovaleric and methylbutyric acid [33]. These acids are fermentation products of valine, leucine and isoleucine [112].

Relative rates of deamination of amino acids by washed suspensions of rumen bacteria were observed qualitatively [60]. Deamination of proline to δ-amino valeric acid by washed suspension was only possible in the presence of hydrogen-donating amino acids, e.g. alanine [60]. Lewis and Emery [98] showed that proline and hydroxy proline were among the slightly deaminated amino acids in experiments with washed cell suspensions of mixed rumen bacteria exposed to single amino acids, using whole rumen contents, deamination of proline alone was possible as many hydrogen donors are available [11]. Hungate [85] reviewed amino acid fermentation in the rumen quite amply.

More recently, it has been proposed that proteolysis and not deamination is the rate-limiting step [125]. However, both operations, i.e. proteolysis and deamination, are affected by pH. Below pH 4.5 and higher than pH 7.2, deamination stops [35, 98].

Proteolysis cannot yield energy-rich ATP components, but further degradation by deamination and subsequent decarboxylation gave 1–8 mol of ATP [51] from 100 g of casein. Protein degradation rates are affected by many factors which were discussed by Tamminga [152]. These factors could be summarised in the following: (1) Solubility of the protein which correlates with the protein degradation rate. (2) Influence of pH on protein solubility will affect protein degradation [87]. (3) Treatments during manufacturing especially with forages and oil cakes, e.g. drying, decrease nitrogen solubility [149]. (4) Weather conditions, e.g. rain, removes soluble nitrogen from forages. (5) Factors related to the animals can also affect the rate of degradation, i.e. animal species, age, condition. (6) Retention time affects the rate of proteolysis and protein breakdown in the rumen. (7) Level of feed intake was shown to influence protein breakdown in the rumen. Higher levels of feed intake depressed the degradation rate [153].

At a low level of intake, 26% of the total dietary N escaped microbial degradation in the rumen. At higher levels of intake, 42% escaped degradation.

C. Utilisation of NH_3-N by Rumen Micro-Organisms

1. Factors Affecting Urea Utilisation by Rumen Micro-Organisms

Ammonia-N is utilized by rumen bacteria for the synthesis of microbial protein. Several factors are necessary for best utilization.

(a) Rate of Ammonia Release in the Rumen. It was suggested that for best urea-N utilization urea should be evenly distributed over a 24-hour

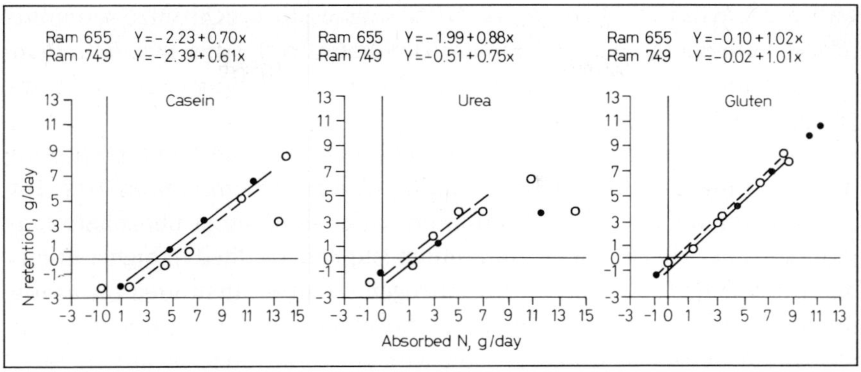

Fig. 2. Relationship between absorbed nitrogen and N retention in rams No. 655 and 749 receiving diets containing casein, urea or gluten, with regression equations for the relationship in each ram.

period. However, there was no difference in urea-N utilization if it was fed twice daily or continuously infused into the rumen [148]. The rate of ammonia release should be optimum for the rate of utilization by rumen bacteria [127]. Deif et al. [50] found that dividing the urea-N in the ration into three equal portions caused better N balance than feeding it in one portion. In a second experiment on sheep, Deif et al. [49] used three sources of nitrogen, urea (NPN) casein (soluble protein) and rice gluten (insoluble protein), and determined the biological value employing the technique of Allison and Anderson [7]. Rice gluten had a biological value of 1.05, urea at a level of 5 g n gave a biological value (BV) of 0.82. When the level increased to 10 g N the BV was depressed to 0.3. Casein had a biological value of 0.65 (fig. 2). From this experiment is was concluded: (A) As urea-N increased in the ration over 5 g, the BV and hence the urea-N utilization was depressed. (B) Rice gluten meal (a product of the Starch and Yeast Co., Alexandria) had a BV of approximately 100% (K′ = 1), hence according to the equation: U 0 (1+K) A + U, where U = urinary nitrogen, K = index of NB for absorbed N(A) or (BV) and Uo = endogenous urinary N. Where K = 1, therefore U = Uo, which provides a method for estimating endogenous urinary-N without having to fast the animal nitrogenously.

(b) The presence in the Diet of a Readily Available Source of Carbohydrates. Fonnesbeck et al. [69] wrote ‘NPN compounds that are hydrolyzed to ammonia at a slower rate may be fed with feeds higher in digested cell

wall carbohydrates. Blending of NPN sources to synchronize ammonia release with energy release from carbohydrates in the diet may be a management tool to attain the "optimum" level of ammonia release in any given set of circumstances.'

(c) Low pH Values of the Rumen. This depresses ammonia absorption from the rumen since it will be found in the form of ammonium ions. The use of compounds such as ammonium sulphate or urea phosphate may induce lower pH. The use of ammonium sulphate resulted in higher microbial protein synthesis and greater nitrogen retention than urea furnishing the same nitrogen intake [30]. Urea phosphate did not improve nitrogen utilization by sheep in comparison with urea and bone meal (similar P level for the two rations) [66].

(d) A Well-Balanced Diet with Respect to Minerals. There is good evidence that some elements exert a beneficial effect on urea-N utilization. Important among these is sulphur. A ratio of 10–15:1 of N:S causes better utilization of urea-N [50, 41, 114, 127]. Sulphur is important for the synthesis of sulphur amino acids, methionine and cysteine. More sulphur is needed for sheep and lactating animals because of the higher sulphur amino acid content of wool and milk proteins.

Experiments conducted at Alexandria University showed specific requirements for potassium by ruminants fed on urea. Especially animals having no green forages in their diets. Rams equipped with permanent rumen fistulae were fed a semi-purified diet containing urea as the only source of nitrogen (results were corrected for undigestible protein-N present in wheat straw). Microbial protein-N was determined by precipitation with tungstic acid at the beginning of the experiment then again after 7 days when the animals showed signs of vitamin B_1 (thiamine) deficiency [115].

Microbial protein concentration and rate of synthesis was significantly depressed at the onset of vitamin deficiency symptoms than before their appearance, when the animals were normal. The symptoms were relieved after treatment with vitamin B_1 or B complex. Microbial protein synthesis had slightly improved but was much lower than in normal animals. When potassium was administered, the rate of synthesis returned to normal (table I). Samples were collected from the rumen of sheep exhibiting symptoms of deficiency and incubated in vitro with different levels of K. When the K level reached 20 mg/100 ml of rumen liquor, both bacterial and protozoal protein synthesis improved significantly but started to decline again at higher levels of K (table II) [116]. Juhasz et al. [90] found that

Table I. Experiment 2: concentration and rate of increase of bacterial and protozoal N in digesta from sheep depleted of potassium (mean of 2 sheep)

Nutritional state of the animals	Bacterial		Protozoal N	
	concentration[1] mg/100 g r.c.	growth rate[2] %/h	concentration[1] mg/100 g r.c.	growth rate[2] %/h
Normal	23.5	26.6	60.5	13.1
Deficient	20.6	–22.1	19.3	6.4
After vitamin B treatment	26.6	–11.1	14.8	15.5
Supplementation with potassium	20.2	25.1	14.8	52.1

After Naga et al. [116].
[1] Digesta as taken from rumen.
[2] Growth in vitro.

Table II. Effect of different levels of potassium on the in vitro growth rates (mean ± SE) of bacteria and protozoa obtained from the rumen of a vitamin B-deficient sheep

Level of potassium supplementation mg/100 g r.c.	Growth rate (%/h) of	
	bacteria	protozoa
0	12.4 ± 6.1	40.0 ± 30.7
10	27.0 ± 5.4	16.2 ± 26.7
20	58.2 ± 19.5	35.0 ± 20.0
30	51.6 ± 14.5	19.6 ± 22.5
40	34.6 ± 18.9	17.6 ± 18.1
60	26.3 ± 22.6	45.8 ± 24.0

Results are averages of four replicates, two on each animal. After Naga et al. [116].

increased ammonia in plasma caused higher losses in potassium in the urine which affects bacterial protein synthesis.

Buffalo and cattle bulls were put on fattening rations containing cottonseed cake (CSC) (46%) CSC + urea (11.5 + 1.5%), 2% urea + sulphur, and urea + KCl (0.5%). The experiment lasted for 5 months. No green forages were administered. There was a marked depression in growth rate

Table III. Live weight gain of male and female buffalo and cow calves between the ages of 4 and 9 months fed on different nitrogenous supplements

Animal species and sex	Average live weight gain, kg/day					
	1[1]	2	3	4	5	6
Buffalos						
Male	0.81	0.91	0.71	0.81	0.64	0.95
Female	0.88	0.79	0.75	0.85	0.82	0.91
Cows						
Male	0.78	0.74	0.55	0.71	0.81	0.67
Female	0.94	0.93	0.88	0.83	0.71	0.69

[1] Concentrate mixtures: 1 = 46% cottonseed cake; 2 = 11% cottonseed cake + 1.5% urea; 3 = 2% urea; 4 = 2% urea; 5 = 2% urea + 0.5% sulphate; 6 = 2% urea + 0.5% KCl.

when urea alone was used. With 0.5% KCl there was a significant increase in growth rate (table III) of buffalo calves only.

More studies are required to determine the possible value of other macro- and micro-elements on urea-N utilization.

(e) Inhibition of Urease Activities. Different substances were used to inhibit urease activity in the rumen; barbituric acid [43], S-2 carboxy ethyl 3-thiosulphopropionate [31], and acetohydroxamic acid [147]. These compounds depressed NH_3 production in vitro [100] but their responses in animal performance were variable.

(f) Level and Quality of Proteins. Satter and Slyter [137] examined microbial protein synthesis in vitro by tungstic acid precipitation against protein level and ammonia-N concentration and found that 13% protein is adequate for protein synthesis. Above this level ammonia is probably wasted.

D. Microbial Protein Synthesis

The concept that feed proteins have similar values for ruminants was accepted by many scientists earlier, since it was shown that growing and lactating cattle could produce with no dietary proteins [126, 157]. Johnson et al. [88] and Dyer and Fletcher [55] found small non-significant differences in protein quality of the diets given to ruminants. This was explained on the basis of conversion of dietary proteins into microbial proteins. As mentioned above, dietary proteins are degraded into amino acids and ammonia which are utilized by rumen micro-organisms for the synthesis of microbial proteins [85].

This has changed since it was found that the rates of ammonia production in the rumen are greatly affected by the solubility of proteins. Deamination of amino acids is usually faster than proteolysis resulting in low concentration of amino acids unless deamination is inhibited [13].

Chalmers et al. [36] obtained evidence of improved nitrogen balance in sheep given casein through the duodenal fistula than through the rumen fistula. Also, casein particles treated by heat and stained with different colours than those untreated remained in the rumen of sheep for a longer period [Chelmers and Synge, personal commun.]. These experiments led to a breakthrough in ruminant nutrition since it became possible to improve the utilization of protein, casein and soybeans simply by heat or treatment with formaldehyde which resulted in an increase in wool production [68].

These pioneer results have also led to the use of methionine, one of the limiting amino acids, by introducing it through an abomasal or a duodenal fistula or by protecting the amino acid from being attacked by rumen micro-organisms by introducing it through a specific capsule that is not dissolved in the rumen but is rapidly hydrolysed to release the methionine in the abomasum or the duodenum, where the pH is suitable. Methionine is released and absorbed into the blood where it is utilized by the tissues [53].

These pioneer results have also changed the generally accepted concept that ruminants have no requirements for essential amino acids because the 10 essential amino acids could all be synthesized by rumen micro-organisms. McDonald [107] proved that lysine, the essential amino acid which is absent from zein, was synthesized in the rumen of sheep having zein as the only source of nitrogen in the rumen. Loosli et al. [101] showed that methionine can be synthesized from sulphate (^{35}S), and that all ten essential amino acids could be synthesized from ammonia [8]. This led to the wrong conclusion that there are no essential amino acid requirements for ruminants since all ten are synthesized by micro-organisms from the rumen. While this may be true for low production, high-producing animals showed greater requirements for certain essential amino acids. Infusion of amino acids into the abomasum or duodenum of sheep suggested that methionine, lysine and threonine are limiting amino acids [67, 119, 135]. Other studies suggested that lysine, threonine, histidine and

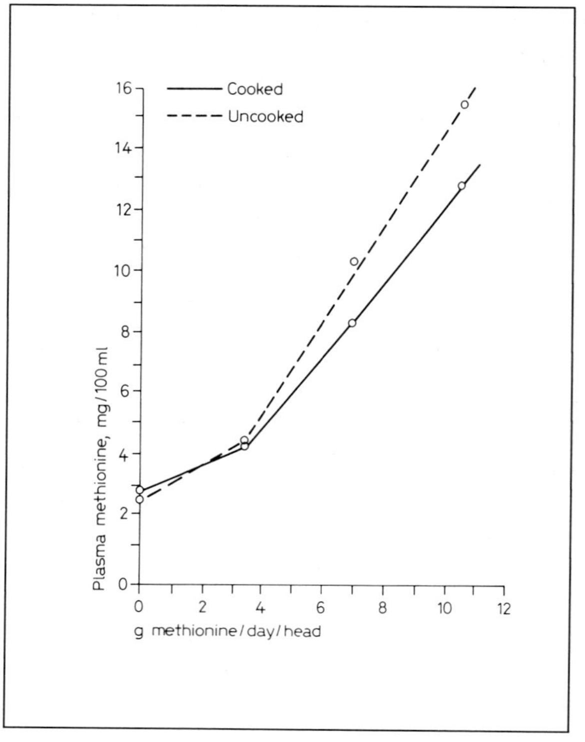

Fig. 3. Free methionine in plasma of early weaned calves given *DL*-methionine intraperitoneally. From Mesbah [113].

phenylalanine may be limiting [38, 77]. Plasma amino acid profiles proved useful in identifying the limiting amino acids in ruminants [17].

Plasma free amino acids concentration remained almost constant with increased levels of infused amino acids as long as there was a need by the tissues for the tested amino acid. As soon as the tissue needs were satisfied, the free amino acid concentration in the plasma rose suddenly. The point of inflection of the two curves represents the amino acid requirement under the set conditions [40].

The same technique was applied by Mesbah [113] on early weaned buffalo calves and found that 3.5 g of methionine injected daily intraperitoneally represented the daily requirements of the buffalo calf in addition to that present in the starter meal (fig. 3). The same results were obtained using nitrogen balance trials (fig. 4).

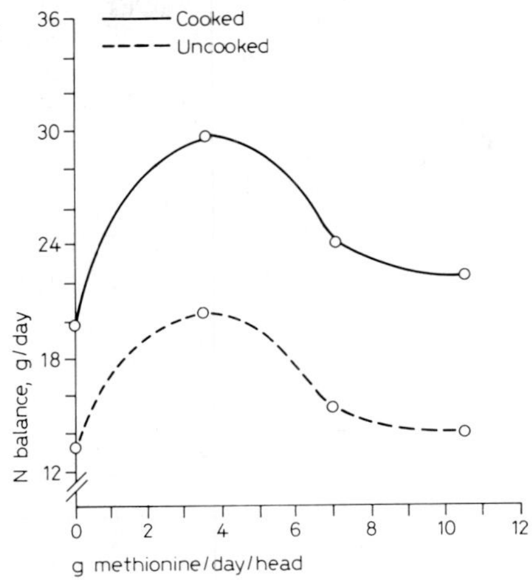

Fig. 4. Nitrogen balance in early weaned buffalo calves given *DL*-methionine intraperitoneally. From Mesbah [113].

By courtesy of Degussa Co. (Konstanz, FRG), we were provided with protected methionine which can bypass the rumen and becomes degraded in the abomasum and the duodenum. This was tested on 70 friesian milking cows in the second season. The cows were chosen according to their first lactation records. Twenty cows giving 3,000–4,000 litres of milk, 30 cows giving 4,000–5,000 litres and 20 cows giving over 5,000 litres per season were used in the experiment. The first group received 10 g, the 2nd group received 20 g and the third group 30 g of Mepron daily. Mepron was added to the concentrate feed administered in the parlour while the cows were milking, starting 20 days before calving and the treatment continued for 100 days after calving. Thes cows were fed at 110% of the NRC recommended requirements [124]. Milk yield was 14% higher for the three groups (fig. 5) with highest difference from the cows giving over 5,000 litres of milk, the difference was 26%. It was also noticed that calves born to the treated mothers were heavier (fig. 6).

Recent experiments [Alexandria University, unpubl. results] showed that early weaned calves given Mepron at 0.5% of the diet, grew at a faster

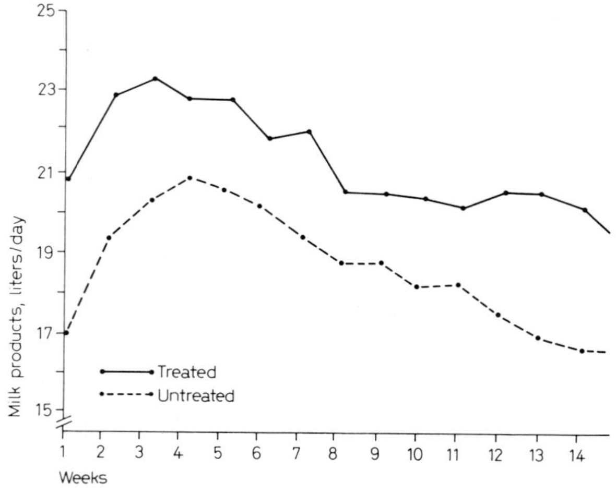

Fig. 5. The effect of adding Mepron (protected methionine) on milk yield [unpubl. results].

rate than the control calves (0.67 vs. 0.52 kg/day in summer and 0.43 vs. 0.25 kg/day in winter). When the starter meal contained soya beans instead of *V. faba* the treated calves made 0.71 kg gain/day vs. 0.47 kg/day for the untreated calves [unpubl. results]. Oke et al. [128] studied the effect of protected methionine and lysine on lambs and steers. The results indicated improvement in average daily gain and N balance.

The tissue requirements for essential amino acids in ruminants and monogastric animals are similar [54]. However, the ruminants depend for the requirement of essential amino acids on microbial protein synthesis in the reticulo-rumen and on amino acids escaping the rumen. These are furnished with (1) the undegraded portion of dietary protein which bypassed the rumen; (2) protected proteins or amino acids, and (3) the resistant amino acids to microbial attack, e.g. methionine, serine, glycine, tyrosine and cystine which were found to be degraded to a lesser extent. Valine and methionine are more resistant to microbial degradation [39].

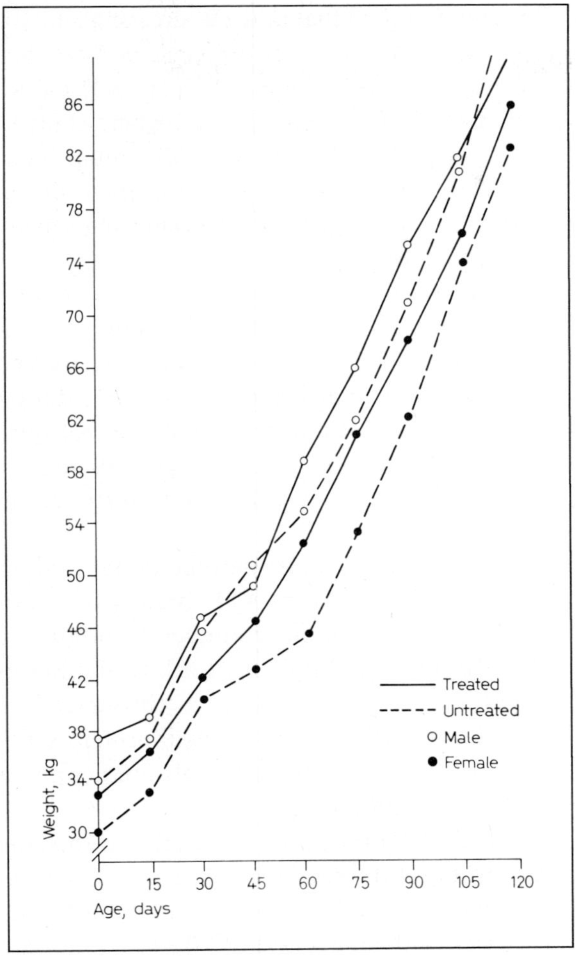

Fig. 6. Effect of supplementation of Mepron (protected methionine) on cows' rations on the growth rate of calves (males and females).

E. Yield of Microbial Protein in Relation to Volatile Fatty Acid Production

Rumen micro-organisms use a portion of the energy released from carbohydrate breakdown to form volatile fatty acids, hence rich energy ATP. For each molecule of volatile fatty acid formed (mixture of acetic, propionic and butyric, acid) 2–4 mol of ATP are produced [79].

Bauchop and Elsden [16] estimated that pure cultures of bacterial cells gave 10 g of dried bacterial cells (DBC) for each molecule of ATP obtained from the VFA formation (Y_{ATP} = 10 g DBC/mol ATP). Hespell and Bryant [76] reported values as high as 15–23 g/mol, but not higher. Many of the Y_{ATP} estimates are in the range of 10–12 g/mol [14]. Maximum Y_{ATP} values were calculated by Hespell and Bryant [76] to be of the order of 26. Factors affecting efficiency of microbial growth are: (a) nutrients availability, and (b) dilution of rumen contents.

This has been shown by El-Shazly and Hungate [62] who estimated the net growth of rumen micro-organisms through the maximum fermentation capacity (maximal rate of gas production in vitro). Zaki el Din and El-Shazly [167] showed that by increasing nutrient availability and dilution of rumen contents, a 3-fold increase in net growth % could be realized. The effect of dilution in improving microbial growth is attributed to the reduced autolysis of bacteria, reduced engulfment by protozoa, and to decreased inhibition [75].

More ATP yield could be achieved under aerobic rather than under anerobic conditions [131]. For effective use of ATP, branched chain fatty acids are important otherwise fermentation continues to produce ATP without being used in the biosynthesis of protein [136]. Under low availability of N microbial polysaccharide formation is encouraged [106]. Recently, evidence that VFA production and microbial protein synthesis go in opposite directions makes it difficult to accurately predict microbial protein synthesis from VFA production [95].

Under the whole range of VFA production, only around the point at which the two curves meet is it that the results can be accurate. Microbial protein synthesis in the rumen varies from 9.6 to 33.2 g/100 g organic matter fermented in the rumen with an average of 20 g [154] or 7.6–20.3 with an average of 12.6 g/100 organic matter (OM). Microbial protein yield (g/100 g OM) is not a constant value. It depends on the method of determination of microbial protein and on the growth rate of micro-organisms which in turn depends on nutrient availability and dilution of rumen contents.

F. Methods for Estimating Microbial Growth and Protein Biosynthesis

The methods have been critically reviewed by Stern and Hoover [144] who also discussed the factors affecting microbial protein synthesis. They reported values ranging from 6.2 g CP/100 g OM digested for sheep fed essentially on lucerne meal + casein using ^{35}S [93], to a high value of 27.5 g CP/100 g OM digested found for sheep fed a straw + urea diet using DAP as marker [80]. Results of numerous studies under varied conditions of feeding using different techniques suggests an approximate average value of 16.9 g microbial protein per 100 g OM digested in the rumen.

A method which was not reported in the review is that of estimating the net growth through measuring the fermentation capacity of the rumen contents in vitro, i.e. maximal rate of gas production which reflects the active biomass of the rumen contents. The increase (or decrease) in the slope of the maximal fermentation curves before and after 1 h of incubation of the rumen samples obtained before or after feeding the animals indicates the increased (or decreased) capacity of the microbiota to ferment the substrate, which is a measure of the growth rate of the microorganisms (positive or negative). The average growth rate constant for all experiments (over a 24-hour cycle) was about 8%/h, which means a turnover rate of 1.92 times/day [62].

Other methods of measuring mcirobial proteins are:

1. The Use of ^{35}S as a Marker of Microbial Proteins. The method gives higher results as part of the radioisotope may have been incorporated into the proteins. Walker and Nader [158] and Walker et al. [159] compared estimates of microbial proteins synthesis using ^{35}S with values using DAP. The latter gave 30% lower values because DAP do not account for protozoal protein.

2. The Use of ^{32}P as a Marker. Radioactive ^{32}P has also been used as a marker for microbial protein synthesis determination. Smith and McAllen [139] and Smith et al. [140] incorporated ^{32}P into rumen bacterial nucleic acids and measured the nucleic acid ^{32}P:NAN ratio in the rumen or duodenal digesta. These estimates were 85% of those based on total RNA.

Other markers used are ATP [70] and ribonucleic acid [103, 104]. Amino acid profiles in the duodenal digesta were also used to estimate microbial and dietary proteins [65].

3. The Estimation of AA Proportion of Bacterial and Protozoal N. More recently, a method for estimating the proportions of bacterial and protozoal N in the total NAN reaching the lower gut of the ruminants under steady state conditions was evaluated [142]. Infusion of ^{15}N-labeled ammonium sulphate for 68 h was undertaken. Bacteria and protozoal samples were isolated at 6-hour intervals. Samples were analysed for ^{15}N enrichment and run through a computer programme. Estimates of 0.39–0.45 and

0.22–0.41 for the proportion of duodenal NAN derived from bacteria and protozoa, respectively, were obtained.

4. The Use of α, ε*-Diamino Pimelic Acid (DAP) and 2-Aminoethyl Phosphonic Acids (AED) as Markers for Bacterial and Protozoal Proteins.* Work and Dewey [164] have shown that the DAP:N ratio varies among different species of bacteria. Synge [151] suggested the possible use of DAP as a marker for rumen bacteria. Weller et al. [161] estimated the rate of synthesis of bacterial protein using DAP. El-Shazly and Hungate [63] developed a method for separating DAP on a Dowex 50 × 8 column (50 cm long) eluting with 6 *N* HCl. The fraction containing DAP also contained other amino acids but DAP was quantitatively estimated using acid-ninhydrin which gave a yellow colour while other contaminating amino acids did not react at this acid pH [165]. Mason and White [102] developed the method further by using a shorter column of 18 cm. Samples of rumen contents before feeding and at different hours after feeding were hydrolysed for 22 h with 6 *N* HCl and put on the Dowex 50 × 8 column for estimating bacterial protein concentration and rate of synthesis in vitro.

After AEP was discovered in ciliate protozoa from the rumen [81], Abou Akkada et al. [3] separated AEP on a Dowex 50 × 8 column and showed that it was not found in feeds or bacteria. El-Shazly et al. [64] used a 10-cm column of Dowex 50 × 8 and eluted with 6 *N* HCl to separate a fraction which contained AEP as well as 6 other amino acids. The rumen samples were hydrolysed for 48 h in 6 *N* HCl, which will break all phosphoric acid ester bonds but not the phosphonic bond. This fraction was put on a one-dimensional filter paper chromatogram to separate the seven amino acid components of this fraction. A second (spot C) amino acid reacted to the phosphorus-Fisk and Subbarow test which showed that there is a second aminophosphonic acid separated with this fraction (fig. 7). This aminophosphonic acid was found in rumen contents, in bacteria and in protozoa. AEP was only found in protozoa. AEP spot on the chromatogram was transferred into a Pyrex test tube where phosphorus was determined.

Using DAP and AEP as markers for bacterial and protozoal synthesis, assuming that bacteria contain 0.57% DAP-N [161] and protozoa contain 1.64% AEP-N [122], microbial protein synthesis was estimated in several trials with sheep, goats, cattle and buffalo. It is essential for every change of diet that a pure fraction of bacteria and protozoa should be separated to determine DAP and AEP concentrations, respectively.

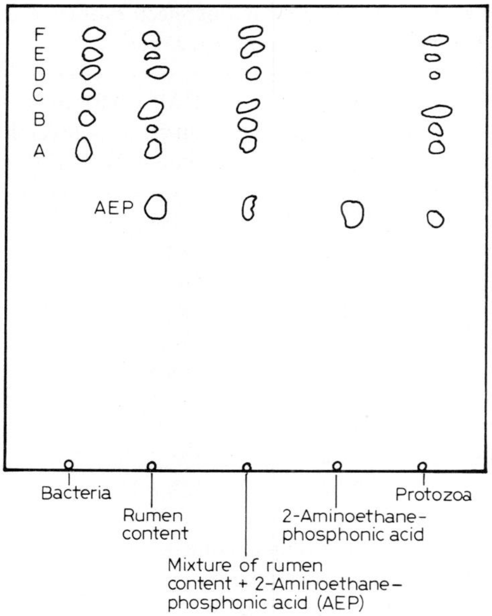

Fig. 7. Single-dimensional chromatogram: solvent, butan-l-ol′-acetic acid-water (8 + 3 + 3). A–F represent six ninhydrin-positive spots. From El-Shazly et al. [64].

The technique was verified by comparison with precipitation of microbial proteins with sodium tungstate. Rumen samples were collected from a ram fed a semipurified diet with urea as the only source of nitrogen. Comparing microbial protein predicted from DAP and AEP with precipitable protein, it represented 97 ± 8.2% (table IV). A similar experiment was repeated on 6 animals, 2 of each of buffalo, sheep and goats, all equipped with rumen fistulae. Microbial protein predicted from DAP + AEP represented 101.77% of the sodium tungstate precipitable protein (table V).

Hydrolysates of cottonseed and rice gluten proteins were added to rumen liquor or rumen contents at 100, 50, 25 and 0% of total N and incubated in vitro for 3 h. Microbial protein-N estimated using DAP + AEP was compared with precipitable protein N. There was excellent corrolation between the two methods. The values were similar for the levels 0, 50 and 75% urea-N but was significantly depressed at 100% urea-N [57].

Table IV. Concentration of microbial protein in the rumen of sheep estimated by two different methods on the third week after feeding a semi-purified diet[1]

Time h	Predicted microbial N			Tungstate A precipitable N mg/100 g r.c.	DAP + AEP predicted N tungstate A precipitable protein-N, × 100
	bacteria[2]	protein[3]	total		
BF	137.6	40.3	177.9	136.4	130.4
3/4	109.9	43.8	155.7	134.8	114.0
1	130.6	20.6	151.2	148.8	101.9
2	91.8	31.3	123.1	151.8	81.1
3	71.1	61.3	132.4	170.3	77.7
4	57.6	66.3	123.9	180.8	68.5
6	109.6	22.5	132.1	185.8	71.1
7	113.0	19.4	132.4	188.8	70.1
10	61.5	64.4	125.9	99.4	126.7
11	59.2	75.0	134.2	104.4	128.5

[1] Corrected for N in wheat straw (42.2 mg N/100 g rumen) contents.
[2] Estimated using the value 0.57% DAP-N of total bacterial N.
[3] Estimated using the value 1.64% AEP-N of total protozoal N.

Table V. Microbial protein concentration in the rumen of sheep, goats and buffaloes fed on a semi-purified diet

Animal species	Microbial protein syn.			Tungstate A precipitable N mg/100 g r.c.	DAP + AEP predicted N tungstate A precipitable protein-N, × 100
	bacteria	protein	total		
Sheep					
a	23.49	60.57	84.06	86.98	96.64
b	28.05	54.68	82.73	80.27	103.06
Goats					
a	25.88	61.01	86.89	84.68	102.61
b	29.99	50.79	80.78	77.44	104.31
Buffaloes					
a	24.82	64.58	89.40	88.79	100.68
b	25.75	81.48	107.23	103.70	103.40
Mean	26.33	62.18	88.51	86.97	101.77

a = Results of accumulative samples of rumen contents obtained from each animal at 7 points after feeding. b = Same results after correction for undigested protein-N in wheat straw.

5. *Comparing DAP + AEP and ^{15}N for Measuring Rumen Microbial Proteins in Bulls* [4]. The rate of incorporation of ^{15}N in $(NH_4)_2SO_4$ enriched with 30% ^{15}N into microbial protein following the method of Al Rabat et al. [10] was found to be 2.3 mg NH_3/100 g rumen contents/h. which could be calculated to yield 59.34 g microbial N/day assuming that 71% of the microbial protein is formed from NH_3 and 29% from amino acids [132]. In the same samples using DAP + AEP as markers of bacterial and protozoal proteins, the yield of total microbial proteins was found to be 50.58 g/day. Harmeyer and Martens [71] claimed that the use of isotopes gives higher values of microbial proteins.

6. *The Yield of Microbial Protein in the Duodenal Digesta.* A ram equipped with two duodenal cannulae, one proximal to the pyloric sphincter and the other 5 cm away from the first [4, 28], was used to compare microbial protein yield as estimated by DAP + AEP and by precipitation with sodium tungstate. The ram was fed a semi-purified diet containing urea as the only source of nitrogen. The duodenal path was blocked by an inflated baloon, inserted in the second cannula.

Duodenal digesta was collected through the first cannula for 12 h. Microbial protein-N was found to be 4.3 and 4.19 g/day for DAP + AEP and sodium tungstate precipitable protein, respectively [26].

In a second experiment using sheep, protozoal + bacterial N in the duodenal digesta as estimated by DAP + AEP was found to be 6.21 g/day while NAN was 5.99 g/day. Urea utilization was 43% [4].

All these results indicated the validity of using DAP and AEP for predicting microbial protein within a reasonable degree of accuracy. Ling and Buttery [99] assessed the use of RNA, ^{35}S, DAP and AEP as markers of rumen microbial proteins. They concluded that the use of AEP as a protozoal marker was invalid because they could find AEP in feeds. Abou Akkada et al. [3], Ibrahim and Ingall [86], and Gzerkawaski [47, 48] could not find AEP in feeds. In our experience using the technique described we could not find AEP in any fraction but protozoa. Another aminophosphonic acid is, however, present in all fractions of rumen contents [64].

Stern and Hoover [144], wrote 'Sampling of rumen contents and the subsequent fractionation of the rumen contents into microbial fractions is one of the most difficult aspects of microbial investigations in the rumen.' In the methods reported above, rumen samples were collected by obtaining 2/3 solid particles and 1/3 liquid which give samples with similar dry matter as that of the rumen contents. These samples are dried and hydrolysed for the determination of DAP and AEP [62].

The method was applied to 4 species of animals: sheep, goats, cattle and buffalo given rations containing 0, 30, 60 and 100% urea-N. The results indicated depression of bacterial and protozoal rates of growth as the level of urea increased in the ration [123].

G. Factors Affecting Microbial Protein Synthesis in the Rumen

As mentioned before, microbial protein synthesis varies greatly from 6.3 to 30.7 g/100 g organic matter digested (DOM). Stern and Hoover [144] reviewed most of the factors affecting microbial protein synthesis which include concentration and source of N, carbohydrates, rumen dilution rates, dietary sulphur and frequency of feeding.

1. *Nitrogen Concentration.* Results of different authors regarding N concentration varied. However, they all agree that there is a certain level over which microbial protein synthesis does not increase under the specified energy level. Many authors suggested a level of 5 mg NH_3-N concentration/100 ml of rumen contents for maximal protein synthesis [12, 117, 129, 138].

2. *The Effect of Protein Source on Microbial Synthesis.* Sheep were fed on diets containing 75% urea and 25% of either cottonseed or rice gluten meal. DAP-N was found to be of the order of 0.59–0.61% of the bacterial N fraction separated from the rumen. Results indicated that microbial protein was 9.22, 10.56 and 11.34 g/100 g DOM, respectively. Bacterial protein represented only 7.16, 8.29 and 8.15 g protein/100 g DOM [123].

Hume [84] obtained microbial protein synthesis values of 17.1, 19.8, 23.3 and 22.5 g/100 g DOM from sheep given urea, gelatin, casein and zein, respectively. It is obvious that the extent of degradability of proteins in the rumen can influence microbial protein synthesis.

3. *Carbohydrate Source.* Starch being a readily available source of energy seems to be the most suitable source for microbial protein synthesis [105]. However, the origin of starch can have some influence on the rate of microbial growth [130].

4. *Dilution Rate.* Rumen dilution rate is the proportion of total rumen volume leaving the rumen per hour [73]. The relationship between rates of outflow of rumen fluid, rate of microbial growth and VFA production and absorption were discussed by Abou Akkada and El-Shazly [14]. There is good agreement between rate of outflow of rumen contents as measured by different markers, e.g. chromium ethylene diamine tetraacetic acid or polyethylene glycol, and microbial growth as measured by net growth, TCA or tungstate-precipitable N.

H. Effect of the Presence or Absence of Protozoa on Nitrogen Metabolism and Microbial Protein Synthesis in the Rumen

Pounden and Hibbs [134] studied the effect of defaunation using copper sulphate as a defaunating agent on calves. They observed certain changes in the appearance of the animals (pot-bellied, rough hair) and on rumen microbial activities (VFA, ammonia). They concluded that the absence of protozoa does not affect live weight gain of the animals.

Abou Akkada and El-Shazly [1] separated newly born lambs from their mothers at birth and kept them in seclusion so as to remain unfaunated. A few weeks later, half the lambs were inoculated with rumen contents from older fistulated animals until a normal population of ciliate protozoa was established. The two groups were fed and treated similarly until the end of the experiment. The results indicated 17% improvement in growth rate of the reinocluted (faunated) animals. Rumen microbial activities were different in the two groups with higher VFA and NH_3 curves for the faunated group. Nitrogen balance was more positive for the inoculated animals but there was higher plasma urea-N in them [2].

Similar results were obtained by Christiansen et al. [42] who reported that faunated lambs showed 27% improvement in average daily gains. Defaunation was done by $CuSO_4$ in half of the lambs.

Eadie and Gill [56] carried out a similar experiment on sheep fed roughage and concentrate diet. The experiment was extended for 1 whole year. They concluded that there were no differences in performance between the two groups at the end of the year. In all these experiments, the diets were characterized by being rich in energy and nitrogen contents.

Knowing that protozoa can consume and digest bacteria in large numbers [44] and that protozoa are being retained in the rumen [162], it is possible that the presence of large numbers of protozoa in the rumen may lower the growth rate and pool size of bacteria through recycling of NH_3-N to bacterial N to NH_3-N in the rumen [120]. Leng [94] suggested that 'the presence of large numbers of protozoa may increase dietary amino acids requirements of ruminant'.

Demeyer et al. [52] tested several low quality roughage supplemented diets on faunated and defaunated sheep and obtained 17–24% improvement in live weight gains in favour of defaunated animals. Bird and Leng [21] and Bird et al. [22] reported increases in body weight gain of defaunated animals (9%) and a large increase in wool growth (37%). The results indicated increased protein availability to intestinal digestion in defaunated lambs. Wool growth is highly responsive to increased amino acid absorption. Wool growth and body weight gain of defaunated sheep grazing pasture showed significant increases indicating a lower balance of amino acids relative to VFA energy in faunated animals [23, 24]. The dynamics of protozoa were studied in the rumen of sheep by labelling protozoa with ^{14}C-choline and returning the labelled cells into the rumen [95]. The author concluded that protozoa reproduce slowly and that 65% of the protozoa die and are degraded in the rumen, only 15% flow to the lower digestive tract. In the absence of rumen protozoa approximately 20% more microbial protein is available for digestion.

In an experiment on early weaned buffalo calves, Borhami et al. [29] found that early establishment of ciliate protozoa induced by inoculation with rumen contents from large animals stimulated calf growth rate only when the feed was of lower quality. Further work is urged for establishing the factors affecting the conditions where protozoa may have a beneficial effect and where ruminants could better perform in their presence or absence. Such experiments should be accompanied by knowledge of energy level and source, rates of outflow of liquid and solid portions of the rumen contents, microbial protein synthesis, and protein degradability.

I. Biological Value of Feed and Microbial Proteins for Ruminants

Since the identification of proteins of low degradability from those which are highly degradable in the rumen, it has become evident that not all proteins have the same biological value for ruminants. The final combination of dietary non-degradable proteins and microbial proteins reaching the abomasum and the duodenum determine the value of the proteins and NPN and their proportion in the diet. It is the final amino acids mixtures in the intestine ready for absorption that give the proteins and/or NPN their ultimate value for ruminants.

All amino acids, including the ten essential amino acids, are synthesised in the rumen by rumen microorganisms [101]. Hence, the biological value of microbial crude protein was shown to vary from 66 to 87% [111, 160]. Bergen et al. [18, 19] confirmed the above values and found the BV, true digestibility and net protein utilization to be 85, 75 and 63 for ruminal bacterial proteins and 82, 87 and 71 for ruminal protozoa. These values do not exclude nucleic acids which, if subtracted from dietary N, the biological value of the true proteins would reach 100%.

Ruminant tissues were shown to have similar requirements of essential amino acids as those of non-ruminants [54].

The biological values reported above were estimated using rats or mice. It is questionable whether such estimates apply to ruminants. Storm and Orskov [145] determined the biological value of rumen microbial preparations by infusion into the abomasum of lambs using the regression of N balance against absorbed N [6, 49] giving values ranging from 61 to 66. Net protein utilization was 53 [146]. Zahra and El-Shazly [166] estimated the biological value for a number of proteins by regression against absorbed N and found values of 71, 73, 71 and 80 for cottonseed meal, linseed meal, faba beans and sesame meal, respectively. They differed from those estimated by the equation: $BV = A - (U - U_0)/A$ which gave values of 55, 75 and 75 for cottonseed meal, linseed meal, faba beans and sesame meal, respectively. The BV of cottonseed meal was greatly improved by adding Ca^{++} to the meal. Cottonseed meal is rich in phosphorous.

III. Conclusions

For economic feeding of ruminants, it is recommended to maximize NPN in the diet (urea, biuret, etc.) ascertaining maximal utilization of NPN through: (1) Synchronization of NH_3-N with readily available source of fermentable carbohydrates, e.g. starch. This may require slowing the rate of NH_3-N release by using biuret, starrea, or other complexes of ammonia and carbohydrates. (2) If an excessive dose of urea-N over 5 g N has to be used in the diet of sheep (or the equivalent proportion for cattle) to satisfy the nitrogen requirements, it should be given in more than one dose for best utilization. (3) A balanced mineral supplement essentially of S and K and may be of other minerals.

At the same time, minimize the use of proteins to obtain: (1) the necessary preformed amino acids from a partially degradable type of protein for maximal microbial protein synthesis, and (2) the minimum amount of a protein low in degradability, e.g. fish meal or treated proteins (by heat or formaldehyde), to furnish the animal with a balanced amino acid mixture post-ruminally for best production.

It is therefore suggested for future studies to:

(1) Obtain further information on the limiting amino acid requirements under different physiological conditions especially for the fast-growing periods, late pregnancy and peak of lactation, and under conditions of stress.

(2) Estimate the biological value of different proteins in combination with NPN. At the same time collect information on protein degradability, microbial protein synthesis rates of outflow from the rumen, amino acids in the duodenal digesta and in lower parts of the intestine, and their profile in the plasma,. Rates of VFA production should also be determined.

(3) Examine the relationship of minerals, micro and macro, with protein synthesis of rumen micro-organisms and the biological values of dietary and microbial proteins.

(4) Assess the effect of presence or absence of ciliate protozoa on animal performance and production under different levels of energy and N intakes.

IV. Summary

NPN, as well as degradable proteins, gives rise to ammonia production in the rumen. Ammonia is either absorbed or utilized by the rumen bacteria, which have specific requirements for ammonia-N (80%). For maximum utilization of ammonia-N by rumen bacteria, certain amino acids are necessary as they are taken up by bacterial cells. Factors affecting utilization of NH_3-N include slow release of NH_3 to synchronize with fermentable carbohydrates which provide the necessary ATP for the biosynthesis of the peptide bond. A nitrogen:sulphur ratio of the order of 10–15:1 seems to give the highest microbial N yield. Dividing the dose of urea gives higher BV for urea. Microbial proteins range from 6 to 35 g/100 g DOM, with average values of 10–12 g. The BV of microbial proteins tested on rodents ranges from 66 to 87. Lower values were reported for recent estimates in the ruminant, i.e. 61%. Some of the methods used by different workers are discussed. They may be partly responsible for the wide range of estimated values.

References

1 Abou Akkada, A.R.; El-Shazly, K.: Effect of absence of ciliate protozoa from the rumen on microbial activity and growth of lambs. Appl. Microbiol. *12:* 384–390 (1964).

2 Abou Akkada, A.R.; El-Shazly, K.: Effect of presence or absence of rumen ciliate protozoa on some blood components, nitrogen retention and digestibility of food constituents in lambs. J. agric. Sci. *64:* 251–255 (1965).

3 Abou Akkada, A.R.; Messmer, D.A.; Fina, L.R.; Bartley, E.E.: Distribution of 2-aminoethyl phosphonic acid in some rumen micro-organisms. J. Dairy Sci. *57:* 78–81 (1968).

4 Abou Akkada, A.R.; El-Shazly, K.: Non-protein nitrogen utilization and microbial synthesis in the rumen; in Tracer studies on non-protein nitrogen for ruminants, part III, pp. 49–56 FAO/IAEA, Vienna 1976).

5 Ahmed, M.H.; Farid, M.; Borhami, B.E.; Abaza, M.A.; Nour, A.M.: Urea transfer in ruminants. I. Method of determination of urea transfer in sheep. Z. Tierphysiol. Tierernährg. FuttermittelKde. *45:* 268–274 (1981).

6 Allison, J.B.: Biological evaluation of proteins; in Advances in protein chemistry, pp. 155–196 (Academic Press, London 1949).

7 Allison, J.B.; Anderson, J.A.: The relation between absorbed nitrogen, nitrogen balance and biological value of proteins in adult dogs. J. Nutr. *29:* 413–420 (1945).

8 Allison, M.J.: Biosynthesis of amino acids by ruminal micro-organisms. J. Anim. Sci. *29:* 797–807 (1969).

9 Allison, M.J.: Nitrogen requirements of ruminal bacteria; in Owen, Protein requirements of cattle. Symposium, pp. 128–132, Oklahoma State University Miss. Proc. 109 (1982).

10 Al Rabat, M.F.; Baldwin, R.L.; Wier, W.E.: Microbial growth dependence on amino nitrogen in the bovine rumen. A quantitative study. J. Dairy Sci. *54:* 1162–1171 (1971).

11 Annison, E.F.; Chalmers, M.I.; Marshall, J.B.; Synge, R.L.M.: Ruminal ammonia formation in relation to the protein requirements of sheep. III. Ruminal ammonia formation with various diets. J. agric. Sci. Camb. *44:* 270–276 (1954).

12 Annison, E.F.: Microbial protein synthesis in relation to amino acid requirements, pp. 141–152; in Tracer studies on non-protein nitrogen for ruminants, part II (FAO/IAEA, Vienna 1975).

13 Annison, E.F.; Lewis, D.: Metabolism in the rumen, p. 184 (Methuen, London 1959).

14 Baldwin, R.L.; Allison, M.J.: Rumen metabolism. J. Anim. Sci. *57:* 461–477 (1983).

15 Bartley, E.E.; Davidovich, A.; Barr, G.W.: Griffel, G.W.; Dayton, A.D.; Deyoe, C.W.; Metchel, R.M.: Ammonia toxicity in cattle. Rumen and blood changes associated with toxicity and treatment methods. J. Anim. Sci. *43:* 835–841 (1976).

16 Bauchop, T.; Elsden, S.R.: The growth of micro-organisms in relation to their energy supply. J. gen. Microbiol. *23:* 457–469 (1960).

17 Bergen, W.G.: Free amino acid in blood of ruminants. Physiological and nutritional regulation. J. Anim. Sci. *49:* 1577–1589 (1979).

18 Bergen, W.G.; Purser D.B.; Cline, J.H.: Determination of limiting amino acids of rumen isolated microbial protein fed to rat. J. Dairy Sci. *51:* 1608–1700 (1968).

19 Bergen, W.G.; Purser, D.B.; Cline, J.H.: Effect of ration on the nutritive quality of rumen microbial protein. J. Anim. Sci. *27:* 1497–1501 (1968).

20 Bergen, W.G.; Yokoyama, M.T.: Productive limits to rumen fermentation. J. Anim. Sci. *45:* 573–584 (1977).

21 Bird, W.H.; Leng, R.A.: The effect of defaunation of the rumen on the growth of cattle on low protein high energy diets. Br. J. Nutr. *40:* 163–167 (1978).

22 Bird, S.H.; Hill, M.K.; Leng, R.A.: The effects of defaunation on the rumen growth of lambs on low protein high energy diets. Br. J. Nutr. *42:* 81–87 (1979).
23 Bird, S.H.; Leng, R.A.: In Farreland Pranvohra, Recent advances in animal nutrition in Australia, p. 48 (University of New England Publishing Unit, Armidale 1983).
24 Bird, S.H.; Leng, R.A.: Further studies on the effect of the presence or absence of protozoa in the rumen on liveweight gain and wool growth of sheep. Br. J. Nutr. *52:* 607–611 (1984).
25 Blackburn, J.H.; Hobson, P.N.: Degradation of protein in the rumen of sheep and redistribution of the protein nitrogen after feeding. Br. J. Nutr. *14:* 445–456 (1960).
26 Borhami, B.E.; El-Shazly, K.: Non-protein-N utilisation in ruminant feeding. Int. Symp. Animal Production in the Tropics, Gezira 1981.
27 Borhami, B.E.; Farid, M.; Ahmed, M.H.; El-Shazly, K.; Naga, M.A.: Urea transfer in ruminants. II. Net urea transfer in sheep under semi-arid conditions. Z. Tierphysiol. Tierernährg. FuttermittelKde. *45:* 274–286 (1981).
28 Borhami, B.E.; Johansen, J.: Digestion and duodenal flow of ammonia-treated straw and sodium hydroxide treated straw supplemented with urea, soybean meal or fish viscera silage. Acta agric. scand. *31:* 245–250 (1981).
29 Borhami, B.E.; El-Shazly, K.; Abou Akkada, A.R.; Ahmed, I.A.: Effect of early establishment of ciliate protozoa in the rumen on microbial activity and growth of early weaned buffalo calves. J. Dairy Sci. *50:* 1654–1660 (1967).
30 Borhami, B.E.: Unpubl. results.
31 Brent, B.E.; Richardson, D.: Apparent intracellular nature of rumen urease. Abstr. 59th Ann. Meet. American Society of Animal Science, Reno 1967. J. Anim. Sci. *26:* 883–953 (1967).
32 Broome, A.W.: The use of non-protein nitrogen in animal feeds; in Swan, Lewis, Proc. Univ. Nottingham School Agr. Sec. Nutritional Conf. Freed Manufacture, pp. 92–113 (Churchill, London 1968).
33 Bryant, M.P.; Doetsch, R.H.: Factors necessary for the growth of bacteroids succisogenes in the volatile acid fraction of rumen fluids. Science *120:* 944–945 (1954).
34 Bryant, M.P.; Robinson, I.M.: Apparent incorporation of ammonia and amino acid carbon during growth of selected species of ruminal bacteria. J. Dairy Sci. *46:* 150–154 (1963).
35 Chalmers, M.I.: Nitrogen utilisation for lactation; in Falconer, Lactation, pp. 379–386 (Butterworth, London 1969).
36 Chalmers, M.I.; Cuthbertson, D.P.; Synge, R.L.M.: Ruminal ammonai formation in relation to the protein requirement of sheep. I. Duodenal administration and heat processing factors influencing fate of casein supplements. J. agric. Sci. *44:* 254–262 (1954).
37 Chalmers, M.I.; Jaffray, A.E.; White, F.: Movements of ammonia after intra-ruminal administration of urea or casein. Symposium. Proc. Nutr. Soc. *30:* 1–10 (1971).
38 Chalupa, W.: Rumen by-pass and protection of proteins and amino acids. J. Dairy Sci. *58:* 1198–1218 (1975).
39 Chalupa, W.: Degradation of amino-acids by mixed rumen microbial population. J. Anim. Sci. *43:* 828–834 (1976).

40 Chalupa, W.; Chandler, J.E.: Amino acid nutrition of ruminants, in Tracer studies on non-protein nitrogen for ruminants, pp. 107–117. Proceedings of a panel (FAO/IAEA, Vienna 1972).

41 Chalupa, W.; Oltjen, R.R.; Slyter, L.L.; Dinius, D.A.: Sulphur deficiency and tolerance in bull calves (Abstract). J. Anim. Sci. *33:* 378 (1971).

42 Christiansen, W.M.; Kwashimar, C.R.; Wise Burroughs: Influence of protozoa upon rumen acid production and liveweight gains in lambs. J. Anim. Sci. *24:* 730–734 (1965).

43 Clifford, A.J.; Bour Dette, J.R.; Tilinan, A.D.: Effects of urease inhibitors on sheep fed diets containing urea. J. Anim. Sci. *27:* 1073–1080 (1968).

44 Coleman, G.S.: In McDonald, Warner, Digestion and metabolism in the ruminant, p. 149 (University of New England, Publishing Unit, Armidale 1975).

45 Cook, A.R.: Urease activity in the rumen of sheep and the isolation of ureolytic bacteria. J. gen. Microbiol. *92:* 32–48 (1976).

46 Crawford, R.J.; Hoover, J.R.W.H.; Sniffen, C.J.; Crooker, B.A.: Degradation of feedstuff nitrogen in the rumen VS nitrogen solubility in three solvents. J. Anim. Sci. *46:* 1768–1773 (1978).

47 Czerkawski, J.W.: Reassessment of efficiency of synthesis of microbial matter in the rumen. J. Dairy Sci. *55:* 971–982 (1974).

48 Czerkawski, J.W.: Methods for determing 2-6-diamino pimelic acid and 2-aminoethyl phosnonic acid in gut contents. J. Sci. Food Agric. *25:* 45–52 (1974).

49 Deif, H.I.; El-Shazly, K.; Abou Akkada, A.R.: The biological evaluation of urea, casein and gluten in the diets of sheep. Br. J. Nutr. *22:* 451–460 (1968).

50 Deif, H.I.; Abou Akkada, A.R.; El-Shazly, K.: Some factors affecting the utilization of nitrogen in the diet of sheep. Alex. J. agric. Res. *18:* 19–24 (1970).

51 Demeijer, D.I.: Ein Kwantitateve Studie Van het metabolisme van Pensmaaginhoud; thesis, State University et Ghent, (1976). J. Anim. Sci. *49:* 1615–1629 (1979).

52 Demeyer, D.I.; Van Nevel, C.J.; VanDevoord, G.: The effect of defaunation on the growth of lambs fed three urea containing diets. Arch. Tierernähr. *32:* 595–604 (1982).

53 Digenis, G.A.; Amos, H.E.; Yang, K.; Mitchell, G.E.; Jr.; Little, C.O.; Swintosky, J.V.; Parish, R.C.; Schelling, G.T.; Dietz, E.M.; Tucker, R.E.: Methionine substitues in ruminant nutrition. I. Stability of nitrogenous compounds related to methionine during in vitro incubation with rumen micro-organisms. J. pharmac. Sci. *63:* 744–749 (1974).

54 Downes, A.M.: On the amino acids essential for the tissues of the sheep. Aust. J. biol. Sci. *14:* 254–266 (1961).

55 Dyer, I.A.; Fletcher, D.W.: Effects of meat meal on steer performance and rumen microbial activity. J. Anim. Sci. *17:* 391–397 (1958).

56 Eadie, M.J.; Gill, I.C.: The effect of the absence of rumen ciliate protozoa on growing lambs fed on a roughage concentrate diet. Br. J. Nutr. *26:* 155–167 (1971).

57 El-Kinani, L.; Nour, A.M.; Borhami, B.E.; Naga, M.A.: Effect of urea replacement of cottonseed cake and rice gluten on ruminal microbial activity and nitrogen utilization by sheep. Alex. J. agric. Res. (in press, 1986).

58 El-Shazly, K.: Some studies on protein metabolism in the rumen; PhD thesis, Aberdeen University (1951).

59 El-Shazly, K.: Degradation of protein in the rumen of sheep. I. Some volatile fatty acids including branched chain isomers found in vitro. Biochem. J. *51:* 640–647 (1952).

60 El-Shazly, K.: Degradation of protein in the rumen of sheep. II. 2-Amino acid degradation by washed suspension of rumen bacteria. Biochem. J. *51:* 647–653 (1952).

61 El-Shazly, K.: Studies on the nutritive value of some common Egyptian feeding stuffs. I. Nitrogen retention and ruminal ammonia curves. J. agric. Sci. *51:* 149–156 (1958).

62 El-Shazly, K.; Hungate, R.E.: Fermentation capacity as a measure of net growth of rumen micro-organisms. Appl. Microbiol. *13:* 62–69 (1965).

63 El-Shazly, K.; Hungate, R.E.: Method for measuring diamino-pimelic acid in total rumen contents and its application to the estimation of bacterial growth. Appl. Microbiol. *14:* 27–30 (1966).

64 El-Shazly, K.; Nour, A.M.; Abou Akkada, A.R.: A method for determing 2-amino ethane phosphonic acid in rumen contents. Analyst *100:* 263–268 (1975).

65 Evans, R.A.; Oxford, R.F.E.; Offer, N.W.: A method for estimating the quantities of microbial and dietary proteins flowing in the duodenal digesta of ruminants (Abstract). Proc nutr. Soc. *34:* 65 A (1975).

66 Fahmy, W.G.: Effect of different sources of nitrogen on the utilization of water hyacinth residues; MSc. thesis (1984).

67 Fenderson, C.L.; Bergen, W.C.: An assessment of essential amino acid requirements of growing steers. J. Anim. Sci. *41:* 759–754 (1975).

68 Ferguson, K.A.: The protection of dietary proteins and amino acids against microbial fermentation in the rumen; in McDonald, Warner, Digestion and metabolism in the ruminants, pp. 448–464 (University of New England, New South Wales 1974).

69 Fonnesbeck, P.J.; Kearl, L.C.; Harris, L.E.: Feed grade biuret as a protein replacement for ruminants. A review. J. Anim. Sci. *40:* 1150–1184 (1975).

70 Forsberg, C.W.; Lam, K.: Use of adenosine 5′-triphosphate as an indicator of microbiota biomass in rumen contents. Appl. environ. Microbiol. *33:* 528–537 (1977).

71 Harmeyer, J.; Martens, H.: In Binnerts, Velman, Zonen, Indirect estimation of microbial growth in the rumen, vol. 31, pp. 25–30 (Wageningen 1975).

72 Harmeyer, J.; Martens, H.: Aspects of urea metabolism in ruminants with reference to goat. J. Dairy Sci. *63:* 1707–1728 (1980).

73 Harrison, D.G.; Beever, D.E.; Thomson, D.I.; Osbourn, D.F.: Manipulation of fermentation in the rumen. J. Sci. Food Agric. *27:* 617–629 (1976).

74 Hendrickx, H.; Martin, J.: In vitro study of the nitrogen metabolism in rumen. C.r. Rech. Inst. Rech. Sci. Ind. Agric., Bruxelles *31:* 7–66 (1963).

75 Hespell, R.B.: Efficiency of growth by ruminal bacteria. Fed. Proc. *38:* 2706–2712 (1979).

76 Hespell, R.B.; Bryant, M.P.: Efficiency of rumen microbial growth. Influence of some theoretical and experimental factors in Y ATP. J. Anim. Sci. *49:* 1640–1659 (1979).

77 Hill, G.M.; Boling, J.A.; Brodley, N.W.: Post ruminal lysine and methionine infusion in steers fed a urea supplemented diet adequate in sulphur. J. Dairy Sci. *63:* 1242–1247 (1980).

78 Hobson, P.N.: The microflora of the rumen (Meadowfield Press, Shildon 1976).
79 Hogan, J.P.; Weston, R.H.: Quantitative aspects of microbial protein synthesis in the rumen; in Phillipson, Physiology of digestion and metabolism in the ruminant, pp. 474–485 (Oriel Press, Newcastle upon Tyne 1970).
80 Hogan, J.P.; Weston, R.H.: The utilization of akali-treated straw by sheep. J. agric. Res. *22:* 951–958 (1971).
81 Horiguchi, M.; Kandatsu, M.: Isolation of 2-amino ethane phosphonic acid from rumen protozoa. Nature, Lond. *184:* 901–902 (1959).
82 Houpt, T.R.: Utilization of blood urea in ruminants. Am. J. Physiol. *197:* 115–120 (1959).
83 Houpt, T.R.: Transfer of urea and ammonia to the rumen; in Phillipson, Physiology of digestion and metabolism in the ruminant, pp. 119–131 (Oriel Press, Newcastle upon Tyne 1970).
84 Hume, T.D.: Synthesis of microbial protein in the rumen. III. The effect of dietary protein. Aust. J. agric. Res. *21:* 305–314 (1970).
85 Hungate, R.E.: Conversions of nitrogenous materials, in The rumen and its microbes, pp. 281–330 (Academic Press, New York 1966).
86 Ibrahim, E.A.; Ingall, J.R.: Microbial protein biothesis in the rumen. J. Dairy Sci. *55:* 971–982 (1972).
87 Isaacs, J.; Owens, F.N.: Protein soluble in rumen fluid (Abstract). J. Anim. Sci. *35:* 267 (1972).
88 Johnson, B.C.; Hamilton, T.S.; Robinson, W.B.; Garey, J.C.: On the mechanism of non-protein nitrogen utilization by ruminants. J. Anim. Sci. *3:* 287–289 (1944).
89 Jones, G.A.: Ureolytic rumen bacteria; in Briggs, Urea as protein supplement, pp. 111–114 (Pergamon Press, New York 1967).
90 Juhasz, B.; Szegedi, B.; Keresztes, M.: Effect of dietary nitrogen level on electrolytic and water metabolism in sheep: K^+ metabolism; in Tracer studies on non-protein nitrogen for ruminants, part II, pp. 123–132 (FAO/IAEA, Vienna 1975).
91 Kennedy, P.M.; Milligan, L.P.: The degradation and utilization of endogenous urea in the gastrointestinal tract of ruminants. Can. J. Anim. Sci. *60:* 205–221 (1980).
92 Komarek, R.J.; Leffel, E.C.; Brorsh, W.H.; Mason, K.R.: Technique for establishment of a rumen pouch. J. appl. Physiol. *15:* 181–184 (1960).
93 Leibholz, J.: Nitrogen metabolism in sheep. II. The flow of amino acids into the duodenum from dietary and microbial sources. Austr. J. agric. Res. *23:* 1073–1083 (1972).
94 Leng, R.A.: In Mc-William, Sutherland, Leng, Reviews in rural science, vol. 2, pp. 185 (University of New England Printing Unit, Armidale 1976).
95 Leng, R.A.: Dynamics of protozoa in the rumen of sheep. Br. J. Nutr. *48:* 399–415 (1982).
96 Lewis, D.: Fate of nitrogenous compounds in the rumen; in Lewis, Digestive physiology and nutrition of the ruminant, pp. 127–139 (Butterworths, London 1960).
97 Lewis, D.: The inter-relationships of individual proteins and carbohydrates during fermentation in the rumen of the sheep. II. The fermentation of starch in the presence of proteins and other substances containing nitrogen. J. agric. Sci. *58:* 73–79 (1962).
98 Lewis, T.R.; Emery, R.S.: Intermediate products in the catabolism of amino-acids by rumen organisms. J. Dairy Sci. *45:* 1363–1368 (1962).

99 Ling, J.R.; Buttery, P.J.: The simultaneous use of ribonucleic acid, ^{35}S, 2.6-diaminopimelic acid and 2-amino ethyl phosphonic acid as markers of microbial nitrogen entering the duodenum of sheep. Br. J. Nutr. *35:* 165–179 (1978).

100 Loper, D.C.; Harbers, L.H.; Richardson, D.: Evaluation of rumen urease inhibitors in vitro. Abstr. 59th Ann. Meet. American Society of Animal Science, Reno 1967. J. Anim. Sci. *26:* 883–953 (1967).

101 Loosli, J.K.; Williams, H.H.; Thomas, W.E.; Terris, F.H.; Maynard, L.A.: Synthesis of amino-acids in the rumen. Science 110: 144–152 (1949).

102 Mason, V.C.; White, F.: The digestion of bacterial constituents in the sheep. I. The metabolism of 2.6-Diamino-pimelic acid. J. agric. Sci. *77:* 91–98 (1971).

103 McAllan, A.B.; Smith, R.H.: Degradation of nucleic acid derivatives by rumen bacteria in vitro. Br. J. Nutr. *29:* 464–474 (1973).

104 McAllan, A.B.; Smith, R.H.: Effect of dietary nitrogen source on carbohydrate metabolism in the rumen of the young steer. Br. J. Nutr. *36:* 511–522 (1974).

105 McAllan, A.B.; Smith, R.H.: Inter-relationships between different chemical components in mixed rumen bacteria. J. agric. Sci. *86:* 639–642 (1976).

106 McAllan, A.B.; Smith, R.H.: Some effects of variation in carbohydrate and nitrogen intakes on the chemical composition of mixed rumen bacteria from young steers. Br. J. Nutr. *37:* 55–61 (1977).

107 McDonald, I.W.: The absorption of ammonia from the rumen of the sheep. Biochem. J. *42:* 584–587 (1948).

108 McDonald, I.W.: The extent of conversion of food protein to microbial protein in the rumen of sheep. J. Physiol. *107:* 21 P (1948).

109 McDonald, I.W.: The extent of conversion of food protein to microbial protein in the rumen. Biochem. J. *56:* 120–125 (1954).

110 McDonald, I.W.; Hall, R.J.: The conversion of casein into microbial protein in the rumen of sheep. Biochem. J. *67:* 400–403 (1956).

111 McNaught, M.L.; Owen, E.C.; Henry, K.M.; Kon, S.K.: The utilization of non-protein nitrogen in the bovine rumen. VIII. The nutritive value of the proteins of preparations of dried rumen bacteria, rumen protozoa and brewer's yeast for rats. Biochem. J. *56:* 151–156 (1954).

112 Menahan, L.H.; Schultz, L.H.: Metabolism of leucine and valine within the rumen. J. Dairy Sci. *47:* 1080–1085 (1964).

113 Mesbah, M.M.A.: Increasing feed efficiency of farm animals as a result of destroying antitrypsin activity in starter meal of early weaned buffalo calves; Phd thesis, Alexandria University (1978).

114 Moir, R.J.; Somers, M.; Bray, A.C.: Utilization of dietary sulphur and nitrogen by ruminants. Sulphur Inst. J. *3:* 15–22 (1968).

115 Naga, M.A.; Harmayer, M.M.; Holler, H.; Schaller, K.: Suspected B vitamin deficiency of sheep fed a protein free urea containing purified diet. J. Anim. Sci. *40:* 1192–1198 (1975).

116 Naga, M.A.; Nour, A.M.; Borhami, B.E.; Zaki El Din, M.; Abaza, M.A.; El-Shazly, K.; Abou Akkada, A.R.; Oltjen, R.R.: Effect of potassium on the rumen microorganisms of animals fed on diets containing urea. Trop. Anim. Prod. *3:* 62–68 (1978).

117 Nikolic, J.A.; Jovanovic, M.; Filipovic, R.: Microbial protein synthesis by bovine rumen content in relation to ammonia concentration; Studies on non-protein nitrogen for ruminants, Part II, pp. 43–55 (FAO/IAEA, Vienna 1975).

118 Nikolic, J.A.; Pavlicevic, A.; Zetemski, D.; Nego-Vanovic, D.: Adaptation to diets containing significant amounts of non-protein nitrogen; in Ruckebusch, Thivend, Digestive physiology and metabolism in ruminants, pp. 603–620 (AVI Publishing Co., Westport 1980).
119 Nimrick, K.; Hatfield, E.E.; Kaminski, J.; Owens, F.N.: Quantitative assessment of supplemental amino acid needs for growing lambs fed urea as the sole nitrogen source. J. Nutr. *100:* 1293–1300 (1970).
120 Noland, J.V.; Leng, R.A.: Dynamic aspects of ammonia and urea metabolism in sheep. Br. J. Nutr. *27:* 177–194 (1972).
121 Norton, B.W.; Mackintosh, J.B.; Armstrong, D.G.: Urea synthesis and degradation in sheep given pelleted-grass diets containing flaked barley, Br. J. Nutr. *48:* 249–264 (1982).
122 Nour, A.M.; Abou Akkada, A.R.; El-Shazly, K.; Naga, M.A.; Borhami, B.E.; Abaza, M.A.: Effect of increased level of urea in the diet of ruminal protozoal counts in four ruminant species. J. Anim. Sci. *49:* 1300–1305 (1979).
123 Nour, A.M.; Naga, M.A.; Zaki El-Din, M.; Borhami, B.E.; Abou Akkada, A.R.; El-Shazly, K.; Abaza, M.A.: Effect of replacing cottonseed N against urea N on the concentration and growth of the microbial population in the rumen of sheep, goats, cattle and buffaloes. Z. Tierphysiol. Tierernährg. FuttermittelKde *46:* 14–20 (1981).
124 NRC. Nutrient requirements of domestic animals. Nutrient requirements of dairy cattle (National Research Council, Washington 1970).
125 Nugent, J.M.A.; Mangan, J.L.: Rumen proteolysis of fractionated leaf protein, casein and bovine serum albumin (Abstract). Proc. nutr. Soc. *37:* 48 A (1978).
126 Oltjen, R.R.: Effect of feeding ruminants non-protein nitrogen as the only nitrogen source. J. Anim. Sci. *23:* 673–682 (1969).
127 Oltjen, R.R.: Development and status of NPN usage of ruminants. Proc. Mont. Nutr. Conf. Mont. State University, Bozeman, 1972, vol. 23, pp. 30–36.
128 Oke, B.O.; Loerch, S.C.; Deetz, L.E.: Effects of rumen protected methionine and lysine on ruminant performance and nutrient metabolism. J. Anim. Sci. *62:* 1101–1112 (1986).
129 Okorie, A.U.; Buttery, P.J.; Lewis, D.: Ammonia concentration and protein synthesis in the rumen (Abstract). Proc. nutr. Soc. *36:* 38 A (1977).
130 Orskov, E.R.; Fraser, R.C.; McDonald, I.: Digestion of concentrates in sheep. Effects of rumen fermentation of barley and maize diets on protein digestion. Br. J. Nutr. *26:* 477–486 (1971).
131 Owen, F.N.; Isaacson, H.R.: Ruminal microbial yields. Factors influencing synthesis and bypass. Fed. Proc. *36:* 198–202 (1977).
132 Pilgrim, A.F.: The production of methane and hydrogen by sheep. Aust. J. Sci. Res. B *1:* 130–138 (1948).
133 Potthast, V.; Rahman, K.A.; Pfeffer, E.: Effects of sugar supplementation on the urea metabolism in sheep. Z. Tierphysiol. Tierernährg. FuttermittlKde. *37:* 126–133 (1976).
134 Pounden, W.D.; Hibbs, J.W.: The influence of pasture and rumen inoculation on the estabilishment of certain micro-organisms in the rumen of young dairy calves. J. Am. vet. Med. Ass. *114:* 33–35 (1949).
135 Richardson, C.R.; Hatfield, E.E.: The limiting amino acids in growing cattle. J. Anim. Sci. *46:* 740–745 (1978).

136 Russel, J.B.; Hespell, R.B.: Microbial rumen fermentaion. J. Dairy Sci. *64:* 1153–1160 (1981).

137 Satter, L.D.; Slyter, L.L.: Effect of ammonia concentration on ruminal microbes in vitro (Abstract) J. Anim. Sci. *35:* 237 (1972).

138 Satter, L.D.; Slyter, L.L.: Effect of ammonia concentration on rumen microbial protein production in vitro. Br. J. Nutr. *32:* 199–208 (1974).

139 Smith, R.H.; McAllan, A.B.: Nucleic acid metabolism in the ruminant. II. Formation of microbial nucleic acids in the rumen in relation to the digestion of food nitrogen and the fate of dietary nucleic acids. Br. J. Nutr. *24:* 545–556 (1970).

140 Smith, R.H.; McAllan, A.B.; Hewitt, D.; Lewis, P.E.: Estimation of amount of microbial and dietary nitrogen compounds entering the duodenum of cattle. J. agric. Sci. Camb. *90:* 557–568 (1978).

141 Somers, M.: Factors influencing the secretion of nitrogen in sheep saliva. I. The distribution of nitrogen in the mixed and parotid saliva of sheep. Aust. J. exp. Biol. med. Sci. *39:* 111–122 (1961).

142 Steinhour, W.D.; Stokes, M.R.; Clark, J.H.; Rogers, J.A.; Davis, C.L.: Estimation of the proportion of nonammonia nitrogen reaching the lower gut of the ruminant derived from bacterial and protozoal nitrogen. Br. J. Nutr. *48:* 417–430 (1982).

143 Stangel, H.J.: History of the use of urea in ruminant feeds; in Briggs, Urea as a protein supplement (Pergamon Press, New York 1967).

144 Stern, M.D.; Hoover, W.H.: Methods for determining and factors affecting rumen microbial protein synthesis. A review. J. Anim. Sci. *49:* 1590–1603 (1979).

145 Storm, E.; Orskov, E.R.: Utilisation of rumen bacteria by ruminants. Ann. Rech. Vet. *10:* 294–296 (1979).

146 Storm, E.; Orskov, E.R.: Biological value and digestibility of rumen microbial protein in lambs small intestine. Proc. nutr. Soc. *41:* 78 A (1982).

147 Streeter, C.L.; Oltjen, R.R.; Slyter, L.L.; Fishbein, W.N.: Urea utilization in wethers receiving the urease inhibitors, acetohydroxamic acid. J. Anim. Sci. *29:* 89–93 (1969).

148 Streeter, C.L.; Little, C.O.; Mitchell, G.E., Jr.; Scott, R.A.: Influence of rate of ruminal administration of urea on nitrogen utilization in lambs. J. Anim. Sci. *37:* 796–799 (1973).

149 Sullivan, J.T.: Drying and storing herbage as hay; in Butler, Baily, Chemistry and biochemistry of herbage, p. 1 (Academic Press, London 1973).

150 Sym, E.A.: Hydrolase activity of the content of the caecum of the horse and the rumen in cattle. I. Introduction, general methods and proteolytic activity. Acta biol. exp., Varsovie *12:* 192–210 (1938).

151 Synge, R.L.M.: Note on the occurrence of diamino pimelic acid in some intestinal micro-organisms from farm animals. J. gen. Microbiol. *9:* 407–409 (1953).

152 Tamminga, S.: Protein degradation in the forestomacks of ruminants. J. Anim. Sci. *49:* 1615–1630 (1979).

153 Tamminga, S.; Vander Koelen, C.J.; Van Vuuren, A.M.: Effect of the level of feed intake on nitrogen entering the small intestine of dairy cows. Livestock Prod. Sci. *6:* 255–262 (1979).

154 Thomas, P.C.: Microbial protein synthesis. Proc. nutr. Soc. *32:* 85–91 (1973).

155 Tiwari, A.D.; Owens, F.N.; Garrigus, U.S.: Metabolic pathway of biuret degradation and formation of buiret complex in the rumen. J. Anim. Sci. *37:* 1390–1395 (1973).

156 Utley, P.R.; Bradley, N.W.; Bolling, J.A.: Effect of restricted water intake, nutrient digestibility and nitrogen metabolism in steers. J. Anim. Sci. *31:* 130–135 (1970).
157 Virtanen, A.I.: Milk production of cows on a protein-free feed. Science *153:* 1603–1614 (1966).
158 Walker, D.I.; Nader, C.J.: Method for measuring microbial growth in the rumen content. Appl. Microbiol. *16:* 1124–1131 (1968).
159 Walker, D.J.; Egan, A.R.; Nader, C.J.; Ulyatt, M.J.; Storer, G.B.: Rumen microbial synthesis and proportion of microbial and non-microbial nitrogen flowing to the intestines of sheep. Aust. J. agric. Res. *26:* 699–708.
160 Williams, V.J.; Moir, R.J.: Ruminal flora studies in sheep. III. The influence of different sources of nitrogen upon nitrogen retention. Aust. J. sci. Res. *B4:* 337–350 (1951).
161 Weller, R.A.; Gray, F.V.; Pilgrim, A.F.: The conversion of plant nitrogen to microbial nitrogen in the rumen of the sheep. Br. J. Nutr. *12:* 421–429 (1958).
162 Weller, R.A.; Pilgrim, A.F.: Passage of protozoa and volatile fatty acids from the rumen of sheep and from a continuous in vitro fermentation system. Br. J. Nutr. *32:* 341–351 (1974).
163 Weston, R.H.; Hogan, J.P.: The transfer of nitrogen from the blood to the rumen in sheep. Aust. J. biol. Sci. *20:* 967–973 (1967).
164 Work, E.; Dewey, D.L.: The distribution of diaminopimelic acid among various micro-organisms. J. gen. Microbiol. *9:* 394–406 (1953).
165 Work, E.: Reaction of ninhydrin in acid solution with straight chain amino-acids containing two amino groups and its application to estimation for diaminopimelic acid. Biochem J. *69:* 416–423 (1957).
166 Zahra, S.A.; El-Shazly, K.: The biological value of some plant proteins used as supplements for sheep. Alex. J. agric. Res. *19:* 173–180 (1971).
167 Zaki El Din, M.; El-Shazly, K.: Some factors affecting fermentation capacity and net growth of rumen micro-organisms. Appl. Microbiol. *18:* 313–317 (1969).
168 Zinn, R.A.; Bull, L.S.; Hemken, R.W.: Degradation of supplemented proteins in the rumen. J. Anim. Sci. *52:* 857–863 (1981).
169 Zinn, R.A.; Owens, F.N.: Site of protein digestion in steers. Predictability, technical newsletter. National Renderers Association Inc., No. 140983, p. 8 (Oklahoma Agricultural Experiment Station, Stillwater 1983).

Dr. Khaled El-Shazly, Faculty of Agriculture, Alexandria University, Alexandria (Egypt)

Subject Index